TEACHING INNOVATIONS IN LIPID SCIENCE

TEACHING INNOVATIONS IN LIPID SCIENCE

EDITED BY
Randall J. Weselake

CRC Press is an imprint of the
Taylor & Francis Group, an **informa** business

CRC Press
Taylor & Francis Group
6000 Broken Sound Parkway NW, Suite 300
Boca Raton, FL 33487-2742

First issued in paperback 2019

ISBN-13: 978-0-8493-7369-5 (hbk)
ISBN-13: 978-0-367-38822-5 (pbk)

Library of Congress Cataloging-in-Publication Data

Teaching innovations in lipid science / editor, Randall J. Weselake.
p. ; cm.
"A CRC title."
Includes bibliographical references and index.
ISBN 978-0-8493-7369-5 (alk. paper)
1. Lipids--Study and teaching. I. Weselake, Randall J. II. Title.
[DNLM: 1. Lipids--chemistry. 2. Teaching. QU 18 T253 2008]

QP751.T43 2008
612'.01577071--dc22 2007020183

Visit the Taylor & Francis Web site at
http://www.taylorandfrancis.com

and the CRC Press Web site at
http://www.crcpress.com

Contents

I Strategies for Teaching Lipid Science to the Public, Students and Teachers

II Demonstrations and Experiments in Lipid Science

Preface

There is no uniformly adopted definition for lipids. William W. Christie has defined lipids as "fatty acids and their derivatives and substances related biosynthetically or functionally to these compounds" (http://www.lipidlibrary.co.uklipids/whatlipid/index.htm). Excellent detailed information on the chemistry and biochemistry of lipids and analysis of lipids can be found at the Lipid Library at this website.

Lipids touch our lives in a number of ways, including the improvement of oilseed crops, preparation of the foods and supplements we consume, multiple aspects of cellular function and human nutrition and issues surrounding our health, and as potentially valuable biofuels and biolubricants to solve environmental problems and provide alternatives to our dwindling petrochemical reserves. Lipid science, however, is often characterized as "unexciting," with students' becoming disenchanted from looking at textbook pages containing long hydrocarbon chains with equally unappealing scientific names. At the same time, the public is mystified at the ever-changing and nonuniform commentary about food choices containing lipids and the health effects of lipids in our foods. The impetus for *Teaching Innovations in Lipid Science* stems from a lively poster session on the same topic held at the Annual Meeting and Expo of the American Oil Chemists Society in Kansas City, Missouri in May 2003. In the months following the successful event, I contacted the contributors to the poster session and a few others about putting together a book that featured both strategies and experiments in teaching lipid science.

This book offers peer-reviewed contributions from lipid science specialists from Canada, the United States, the United Kingdom and Hong Kong. Section I focuses on teaching lipid science to the general public, students at various levels of education and instructors of lipid science. The section begins with a chapter that describes some of the barriers that lipid science specialists face in transmitting accurate information to the public. The next two chapters place a strong emphasis on the development and implementation of creative programs that foster an interest in lipid science, particularly at the high school level. Chapter 4 presents the creative problem-solving approaches that Dr. Karen Schaich has implemented in her lipid chemistry course at Rutgers University. Following this, strategies for involving independent study students at the undergraduate level in a range of lipid science projects are discussed. Methods for evaluating these students are presented along with some information on the career paths that they eventually chose. In Chapter 6, Dr. Lawrence Johnson and colleagues of the University of Iowa explain how sample cards can be used to teach undergraduates and graduate students about the processing of oilseeds and cereals. The final two chapters of this first section are mainly generalized accounts of biotechnology and crop improvement, and isoprenoid biochemistry, both of which may serve as useful resources to professors and other undergraduate instructors. The biotechnology chapter places a strong emphasis on the improvement of oilseed crops and provides some tips on explaining DNA science and crop biotechnology to the public.

Section II of *Teaching Innovations in Lipid Science* begins with two chapters featuring simple demonstrations on the physical properties of lipids that should be useful for teaching aspects of lipid science in middle- and high school classrooms. Chapters 11 to 13 present experiments for analyzing lipids in food oils, plasma and milk. It is anticipated that these chapters will provide a valuable resource for designing and offering lipid science labs at both the senior undergraduate and graduate student level. The chapters on lipid analysis include information on thin layer chromatography, gas chromatography and high performance liquid chromatography. In Chapter 14, Dr. Robert Moreau of the U.S. Department of Agriculture describes the use of convenient enzyme test kits for teaching lipid chemistry. Exercises involving one or more of these test kits could potentially be added to a lab course that begins with chromatographic methods for analyzing lipids. The final chapter in the book presents theory and experiments for studying lipid metabolism in the plant organelle known as the plastid in an area that straddles biochemistry and physiology. Drs. Salvatore Sparace and Kathryn Kleppinger-Sparace have extensive experience in working with plant plastids, the organelle that houses the process of fatty acid synthesis. Methods are described for preparing plant plastids, and studying metabolite uptake and pathway analysis. It is anticipated that this final chapter will represent a valuable resource for use in advanced undergraduate and graduate student labs.

Overall, *Teaching Innovations in Lipid Science* addresses lipid education at numerous levels ranging from educating the public to offering exciting experiments in lipid biochemistry to senior undergraduates and graduate students. We hope this book will inspire the lipid educator to use some of the approaches and methods presented in designing new courses or modifying existing courses. Above all, we hope that it will inspire readers to think about how this information could be used to disseminate lipid science knowledge for their specific purposes and serve as a basis for pursuing additional novel avenues of instruction.

About the Editor

Randall J. Weselake is currently a professor and Tier I Canada Research Chair in Agricultural Lipid Biotechnology with the Department of Agricultural, Food and Nutritional Science at the University of Alberta in Edmonton. He is also a guest researcher with the Plant Biotechnology Institute of the National Research Council of Canada in Saskatoon, Saskatchewan. Dr. Weselake received his doctorate in plant biochemistry from the Department of Plant Science at the University of Manitoba in Winnipeg in 1984. Research leading to his doctorate was conducted at the Grain Research Laboratory of the Canadian Grain Commission, also in Winnipeg. Although trained as a cereal biochemist, he became involved in plant lipid biochemistry when he joined the Plant Biotechnology Institute as a research associate in 1987. Two years later, he joined the Department of Chemistry and Biochemistry at the University of Lethbridge (Lethbridge, Alberta, Canada) where he continued to work in the area of lipid biochemistry for about 15 years. Dr. Weselake served as chair of the department from 1996–1999 and held a University of Lethbridge Board of Governors Research Chair from 2002 until he left for the University of Alberta in 2004. From 1999–2000, he was a visiting scholar in the Department of Biological Sciences at the University of Calgary in Alberta. Currently, at the University of Alberta, he oversees the research activities of eight graduate students, two postdoctoral researchers and a number of technical personnel. Dr. Weselake has designed a course in lipid science that is offered at both the undergraduate and graduate level.

A large component of the projects and programs led by Dr. Weselake focuses on the development of molecular strategies to enhance seed oil content in canola and modify the fatty acid composition of oil from canola and flax. He has also been involved in the development of biochemical and molecular markers for the marbling trait (intramuscular fat) in beef cattle, and investigations of conjugated linoleic acid in milk fat from dairy cattle and the effect of these fatty acid isomers on milk fat synthesis. Dr. Weselake collaborates extensively with other researchers in Canada, the United States and Europe. His research is supported bv provincial and national funding agencies, and, recently, his work has begun to attract the interest of major industries in the plant biotechnology area. Currently, he is serving as the co-leader of the large-scale functional genomics project, "Designing Oilseeds for Tomorrow's Markets," which is funded by Genome Canada, Genome Alberta and partners. Recently, he became the leader of the large-scale "Bioactive Oils Program" funded by AVAC Ltd. (Alberta) and partners. Dr. Weselake is the author of 87 publications in peer-reviewed journals and book chapters, and has delivered about 50 invited presentations at various conferences and other venues. Within the last eight years, he has published eight invited reviews dealing with various aspects of storage lipid synthesis in plants. He has been serving as an associate editor of the journal *Lipids* since 2005.

Currently, Dr. Weselake is a member of the American Oil Chemists Society (AOCS), Canadian Section of the American Oil Chemists Society (CAOCS), American Society of Plant Biology and Canadian Society of Plant Physiologists. He has been a member-at-large of the governing board and the Biotechnology Division of the AOCS since 2004. He served as president of the CAOCS from 2003–2005 and was technical chairperson for the Annual Meeting and Expo of the AOCS held in Montréal in May 2002. Dr. Weselake has been involved in a number of committees dealing with the adjudication of grant applications and research programs. This year he served as a member of the Plant Research Panel for Agriculture and Agri-Food Canada and currently serves (since 2005) as a member of the Scientific Committee for the large-scale Green Crop Network funded by the Natural Sciences and Engineering Research Council of Canada.

Contributors

Robert G. Ackman
Food Science Program
Department of Process Engineering and Applied Science
Dalhousie University
Halifax, Nova Scotia

Suzanne M. Budge
Food Science Program
Department of Process Engineering and Applied Science
Dalhousie University
Halifax, Nova Scotia

Mee-Len Chye
School of Biological Sciences
The University of Hong Kong
Pokfulam, Hong Kong

Cristina Cruz-Hernandez
Nutrition Research Division
Health Canada
Ottawa, Ontario
and
Nutrient Bioavailability
Nestle Research Center
Lausanne, Switzerland

James K. Daun
Agri-Analytical Consulting
Winnipeg, Manitoba

Tara Furukawa-Stoffer
Department of Agricultural, Food and Nutritional Science
University of Alberta
Edmonton, Alberta

David Hildebrand
University of Kentucky
Lexington, Kentucky

Anthony H.C. Huang
College of Natural and Agricultural Sciences
University of California
Riverside, California

Darren Jarboe
Center for Crops Utilization Research
Iowa State University
Ames, Iowa

Lawrence A. Johnson
Center for Crops Utilization Research
Iowa State University
Ames, Iowa

E. Chris Kazala
Department of Agricultural, Food and Nutritional Science
University of Alberta
Edmonton, Alberta

Kathryn F. Kleppinger-Sparace
Department of Genetics and Biochemistry
Clemson University
Clemson, South Carolina

John K.G. Kramer
Food Research Program
Agriculture and Agri-Food Canada
Guelph, Ontario

Arnis Kuksis
Banting and Best Department of Medical Research and Department of Biochemistry
University of Toronto
Toronto, Ontario

Norm Lee
The Manitoba Network for Science and Technology
Manitoba Energy, Science and Technology
Winnipeg, Manitoba

Jeni Maiers
Center for Crops Utilization Research
Iowa State University
Ames, Iowa

Thomas A. McKeon
U.S. Department of Agriculture
Agricultural Research Service
Western Regional Research Center
Albany, California

Robert A. Moreau
U.S. Department of Agriculture
Agricultural Research Service
Eastern Regional Research Center
Wyndmoor, Pennsylvania

Denis J. Murphy
Biotechnology Unit and Research
Division of Biological Sciences
University of Glamorgan
Treforest, United Kingdom

Roman Przybylski
Department of Chemistry and Biochemistry
University of Lethbridge
Lethbridge, Alberta

Karen M. Schaich
Department of Food Science
Rutgers University
New Brunswick, New Jersey

Crystal L. Snyder
Department of Agricultural, Food and Nutritional Science
University of Alberta
Edmonton, Alberta

Salvatore A. Sparace
Department of Biological Sciences
Clemson University
Clemson, South Carolina

Anne Timmins
Department of Food Science and Technology
Dalhousie University
Halifax, Nova Scotia

Randall J. Weselake
Department of Agricultural, Food and Nutritional Science
University of Alberta
Edmonton, Alberta

Section I

Strategies for Teaching Lipid Science to the Public, Students and Teachers

1 Educating the Public about Lipids

Anthony H.C. Huang and Randall J. Weselake*

CONTENTS

1.1 CURRENT AVENUES FOR TRANSMITTING LIPID KNOWLEDGE TO THE PUBLIC

It is an opportune time for lipid researchers and educators. Concerns about lipid consumption and the role of lipids in health are foremost in the minds of the public. In the U.S., Canada and other developed countries, obesity and being overweight have become major health problems. Coronary artery and peripheral vascular diseases are the major causes of death. The problem not only impinges on physical health, but also induces stigmatization and discrimination in society. Obesity and being overweight have also spread to many fast-developing countries, including China and Brazil, as their populations become more affluent. In the U.S., in response to consumer interest and new advances in lipid research, the U.S. Food and Drug Administration (USFDA) continues to implement new regulations for labeling food products, especially describing the content and type of lipid in food. "Fast food" and traditional restaurants, in particular franchises, have revised their menus to become more attractive to health-conscious consumers, with many of their menus revealing the content and type of lipids in each item. Consumers are reading food labels more carefully than ever before and deciding on what they should purchase and eat in accordance with what they perceive as good for their health. Topping the current list of interests are concerns about *trans* fatty acids, percentage saturation, *omega*-3 fatty acids, fat-free products and low-fat products. Some people readily admit that they know very little about lipids and have acquired the ideas of "good" or "bad"

* Corresponding author.

fats from friends or heard or read by chance information from a certain source. Others, convinced they are knowledgeable because of the substantial efforts spent obtaining information from the media, can proclaim this or that food product to be good or bad because of its content of a certain fat. They often sound like experts, but are they?

Unfortunately, the public is far from being expert, or for that matter, well informed about lipids in foods. Most of the public lacks a background in science, let alone chemistry and biology. Whatever science they learned in grade school or college may have little retention in their memories. Those with more curious minds pay attention to what they read, hear and watch. Most look for information on food and health from the media. A 1999 survey by the American Dietetic Association revealed that the public receives information about food and health primarily from the media, such as television and magazines, and secondarily from dietitians and medical doctors.[1] Diligent people try to read extensively and likely encounter a large volume of similar information, but are eventually indoctrinated without knowing that the various items of information might all have come from a single syndicated media source.

Media reporters are well aware of the public's interest and try to convey scientific information as best as they can. Although not always intended, they play a major role in our health, because they strongly influence what we want, and in turn, what the food industry advertises and delivers. Most reporters, however, lack an in-depth knowledge of chemical and biological sciences. Many reporters probably majored in the humanities in college, for a training that sharpened their broad social knowledge, writing skills and ability to disseminate information. We may be asking too much of a reporter to comprehend the details of scientific issues, especially controversial subjects such as fats and health. Thus, media reports are often easily misunderstood by the public. In a situation less known to the outside world, reporters often face internal competition within the media organization for their reports to be chosen for use in a prestigious segment of the media. As a consequence, they write simplified explanations and attempt to make the subjects such as fats and health exciting and pertinent to the public. Often, the information has been oversimplified to the point of being incorrect or even more difficult to comprehend. Minor points raised in the interview of a scientist sometimes become overemphasized or exaggerated in the media article, and thus the report distorts the information the scientist had intended to transmit to the public. Some reporters, especially those in high-profile media, are more conscientious or are required to have reports checked by experts, which include dietitians, doctors, industrial scientists and university professors.

Dietitians and medical doctors are the main sources reporters consult for expert health/dietetic information. They are also the most direct source, ranked after the media, of information for the general public.[1] Most dietitians probably read extensively, but may lack an in-depth knowledge of scientific concepts that would allow them to discriminate among confusing and biased media information and scientific reports (e.g., results of short-term or statistically questionable studies). Although medical doctors receive substantially more professional training than dietitians, their exposure, except for some specialists, to human nutrition is limited. In the U.S.

and Canada, most patients first encounter their family doctors, the "gatekeepers," before seeing specialists. In general, primary care doctors are trained to comprehend broad personal health situations, but not the "nuts and bolts" of specific fields such as fats and health, and especially not the latest advances in the fields. Their focus is on diagnosing and treating diseases with technical approaches. How many primary care doctors know the structure of an omega-3 fatty acid, let alone know the basis for this fatty acid nomenclature? Most primary care doctors are overworked and have little time to carefully examine issues about food (fats) and health and actually acquire the information from the media, just as their patients do. When asked about a certain fat in relation to health, primary care doctors either try to answer the question authoritatively or refer the patient to a dietitian. Many patients are therefore convinced they were properly informed by an expert. Ironically, most commercials on fats or drugs, as legal protection from claims, routinely urge potential consumers to ask a doctor's advice.

Scientists in the lipid field are often asked by reporters about lipids and health in depth. Some provide sound advice while others, while knowledgeable, could be biased in their statements or have a conflict of interest. A 2003 survey by the Health and Nutrition Division of the American Oil Chemists Society (AOCS) asked division members about their preferences for lipid-containing foods.[2] A high percentage of these experts recommended the food that they were most familiar with or were doing research on.

1.2 INTERWOVEN NETWORKS OF INFORMATION AND MISINFORMATION

Overall, the public is educated about fats in food products from an interwoven network of uninformed or underinformed media reporters, medical professionals, potentially biased scientists and politically influenced government agencies. These sources together generate an apparent consensus, all of which might have come from one or a few primary sources. Some of the claims are valid, whereas the others are debatable or oversimplified. On top of these sources are biased and ubiquitous commercials whose primary objective is to sell products. Added to the commercials are the food labels, which are government mandated.[3] Many food companies try to present the information in creative ways and exploit loopholes in regulations. For example, a high-lipid coffee cream is labeled as "fat free" only because the manufacturer cuts down the suggested serving size, so that the lipid content per serving becomes less than 0.5 g, which can then be rounded to 0.

Although general trends on the effects of the fat composition of food on health have emerged in the scientific literature,[4] lipid research carried out by different groups around the world often produces results on the effects of lipids on animal and human health that may not be in agreement. In addition, new findings in lipid nutrition are often accompanied by press releases, thereby providing new information for the media. Thus, the public becomes even more confused about the effects of lipids on health. It is important that an individual does not tailor his or her lipid intake based on a single media report about a new finding, but that the individual is encouraged to seek out further information from other sources.

1.3 LACK OF PUBLIC AWARENESS OF OTHER ASPECTS OF LIPIDS

Currently, in food, most attention has been placed on fats and oils that are either a major ingredient or have a potential effect on health. The public is not keenly aware of other lipids in food products. These lipids could be emulsifiers such as lecithins, monoglycerides and diglycerides. As consumers become more sophisticated and read food labels more carefully, they will become aware of the presence of these lipids in food products and wonder what they are, why they are present in food and what they will do to the body. Consumers and media reporters need to be educated on these other lipids and have ready access to comprehensible and accurate sources of information.

Also, the public is usually uninformed that in some aspects, lipids are good for health and food taste. Lipids good for health include polyunsaturated fatty acids (PUFAs) and several fat-soluble vitamins. Lipids promote food taste because they preserve juices and colors, as well as enhance texture and appearance. The public is also unaware that lipids are used extensively in nonfood products. Most people are unacquainted with the abundant lipids in shampoos and detergents, although some are aware of the phosphates in specific detergents. Few people know about lauryl derivatives (always shown as the major ingredients in product labels) and their role in detergents, along with the fact that these compounds are derived from tropical fruits such as palm and coconut. Neither are they aware of the fact that lipids are used to produce cosmetics, lubricants, paint and fuel. With more information, the public would be able to appreciate these lipids and their uses in nonfood products and make conscious decisions about the products they buy and use in their daily lives.

An emerging trend related to lipids around the world is the use of biodiesel as a renewable fuel. Biodiesel fuel is derived from vegetable oils identical to household cooking oils. The use of such fuel is an important social topic; whether it is viable on a long-term basis remains to be seen. Regardless, people need to be educated about the fuel, how it is produced, and the pros and cons of its future.

1.4 A PUBLIC IN NEED OF EDUCATION ABOUT LIPIDS

In a way, we, the "lipid experts," have to share some burden of the blame for the public's confusion about lipids and health. We have not developed terminology that can be easily understood by nonscientists. We have been in the field too long and are insensitive to the complexity of lipid terms. Are fats and fatty acids the same? What about triglycerides and triacylglycerols; diglycerides and monoglycerides; omega-3, omega-6 and PUFAs; and even the words lipids and fats? Could we simplify the lipid terminology to educate the public? Let us turn to another common segment of society with a similar issue. We complain about the medical profession using a complicated terminology that patients do not understand or cannot even pronounce. Why not simplify pylonephritis to kidney infection, and cholecystectomy to gall bladder removal?

Lipid researchers or educators represent a very small proportion of the population that has become interested in lipids. It is both an opportunity and an obligation on our part, and rightly so, to extend our knowledge to the public. This is our social responsibility, especially since we are paid by taxpayers directly (e.g., public college professors) or the general public indirectly through the sale of lipids (e.g., industrial workers). What can we do?

1.5 EDUCATING THE PUBLIC ABOUT LIPIDS AT THE "GRASS-ROOTS" LEVEL

"Grass-roots" activities in disseminating knowledge are always a major, long-term, worthy endeavor. Whenever and wherever possible, we, as lipid researchers and educators, should deliver public speeches with easy-to-understand content. Many garden clubs, religious unions and senior citizen organizations are interested in having scientists give lectures at their gatherings. Scientists who can communicate in simple terms with the public are in demand to give talks about hot topics such as cooking oils and biodiesel fuel. We should not wait to be invited, but instead, should actively seek invitations. One of us had a pleasant experience in giving lectures at the Osher Institute, which is a nationwide nonprofit organization for educating senior citizens.[5] Education is for all curious minds, and such education can then benefit society. Adults who seek further semiformal education tend to be more active in voicing their opinions. Once educated and motivated, they will have the time and enthusiasm to write to government agencies and industry to make their opinions known. Their efforts could induce or force changes, such as closing loopholes in government regulations on food labeling and denouncing false or misleading advertising.

We also need to be engaged in educating grade-school students. Responsible teachers usually welcome scientists to give talks to their classes, especially on timely topics or those related to the students' well-being, such as health. Kids love to learn new things because they are naturally curious or perhaps enjoy going home to boast of their newfound knowledge about hot topics related to fats and health. The latter could induce parents to read more about lipids and the effects of lipids on health.

1.6 A LIPID COURSE FOR REPORTERS AND OTHER PROFESSIONALS

In view of the major impact of the media on the public, we need to educate the media. It would be useful to write to reporters when we recognize that the information in a report about lipids is incorrect or oversimplified. Then, the media will be alerted to the fact that their readers are knowledgeable and will instill a determination on their part to produce articles based on more accurate facts.

Reporters may desire to have a major overhaul of their knowledge of timely topics such as lipids related to health. A week-long, three-day or even weekend course tailor-made for media reporters on lipids, health and related issues could be appealing to them. Dietitians, civic leaders and others may also benefit from a course of this nature. The AOCS has offered many short courses on special aspects of lipids for industrial workers and could offer a special lipid course to reporters, dietitians and other interested professionals. Professors on some campuses have the administrative infrastructure, such as an extension branch, to offer such a course. Promotion of such a course needs advanced, aggressive national advertisements so that reporters and other professionals could become aware of it.

1.7 DEVELOPMENT OF A WEBSITE ON LIPIDS FOR THE GENERAL PUBLIC AND PROFESSIONALS

In this cyber age, it would be useful to produce a website to educate the public on lipids. At present, several websites on food and health sponsored by government agencies, foundations, special interest groups and individuals are available. Examples are http://www.nutrition.gov (by the U.S. Department of Agriculture), http://www.eatright.org (by the American Dietetic Association) and http://ific.org/about/contact.cfm (by the International Food Information Council, a foundation funded by industry). We need an independent website administered by a committee whose members have no involvement with industry as employees or consultants or researchers involving a particular crop with immediate applications. Identification of volunteers to serve on the committee, such as the contributing authors of this book, will not be difficult. The website would not be sponsored by a special interest group and would not contain advertisements. It would have unique characteristics and distinguish itself from other websites as follows:

> The front page of the website would offer a selection of the levels of depth of knowledge for readers in grades 1–6, grades 7–12, college students, adults, senior citizens, media and professionals (dietitians, doctors). The site would focus on lipids rather than foods and thus include lipids as minor food ingredients and describe their nonfood uses. The site would provide descriptions, pictures and a question-and-answer feature; it would also be interactive. Setting up such a website and maintaining it weekly or monthly would not require a large amount of funding. The site could even serve as the basis for a new foundation. The site would not be associated with an existing large website, such as that of the AOCS, so as not to have an association with an organization with heavy industry involvement and thus generate the appearance of bias or conflict of interest. The website could feature a list of contactable lipid experts who have agreed to be included on a voluntary basis. The website could become the "gold standard" for providing accurate lipid knowledge to the public. A better-educated public with unbiased information on lipids would eventually benefit the AOCS and other related organizations.

1.8 TIMELY ACTION

The timing is right. Lipid experts or educators must seize this golden opportunity to offer unbiased expertise to a society that is in desperate need of accurate information about lipids. Society has supported us and now needs our input. Let's rise to the occasion.

ACKNOWLEDGMENTS

The authors thank Laura Heraty and Chris Kazala for their helpful comments on the manuscript. RJW is thankful for support provided by the Natural Sciences and Engineering Research Council of Canada (Discovery Grant) and the Canada Research Chairs Program.

REFERENCES

1. American Dietetic Association 2000. Nutrition and you: Trends 2000 survey. (This information can be found at http://www.eatright.org.) (2000).
2. Anonymous, What the experts eat. *Inform* 14, 116–117 (2003).
3. U.S. Food and Drug Administration. CBER guidances/Guidelines/Points to consider. http://www.fda.gov/cber/guidelines.htm (updated to 2007).
4. Wahrburg, U. (2004). What are the health effects of fat? *Eur. J. Nutr.* 43 Suppl 1:I/6–11 (2004).
5. Huang, A., Planting seeds of knowledge in golden minds. *Am. Soc. Plant Biologists News* 31, 21 (2004).

2 Fats, Genes and Food Using Lipids as a Tool for Science Education and Public Outreach

Denis J. Murphy

CONTENTS

2.1 INTRODUCTION

In this chapter, I will discuss the use of lipid science as both an educational tool for school and university students, and as a useful method to engage the general public in debates on such key topics as healthy foods, the obesity epidemic and genetic engineering. I will examine a series of topics and case studies based on personal experience over more than 30 years as an educator and researcher. During this time, lipids in one guise or another have frequently become important themes of public interest and controversy.

The chapter begins with a brief overview of the importance of lipids, both scientifically as biological molecules and in a sociomedical context in terms of their relevance to human well-being. This is followed by an account of several case studies of lipid-related education or outreach activities to show how lipid science can be used as a valuable tool to improve public knowledge and awareness of important issues. Finally, there is a discussion of some possible future strategies for a more coordinated approach to the use of lipid science to enhance learning experiences and public engagement with important health-related scientific themes.

Much of the focus of this chapter is on plant lipids, which reflects the author's research interests. But plant lipids cannot be separated from other lipids, for example when we ingest plant lipids they become part of our bodies, i.e., they become "human lipids." Therefore, I will use plant lipids as a familiar "springboard" to engage in discussions about other types of lipids, such as those in meat and dairy products and in human tissues.

2.1.1 Terminology: Lipids and Students

Lipids are often equated with acyl derivatives, but strictly speaking they encompass all those biomolecules soluble in organic solvents like chloroform/methanol or acetone (Murphy, 2005). In terms of education and outreach, we have found it useful to use the widest possible definition of a lipid. Hence, we include all the acyl lipids such as the membrane phospho- and glyco-lipids, as well as the storage triacylglycerols of adipose tissue, and also all the lipophilic vitamins (A, D and E), plus the dozens of lipidic hormones and physiological mediators, from eicosanoids and prostaglandins to steroids and steryl esters. In discussing teaching and learning, the term "student" here means a university undergraduate—normally a third-year BSc biologist/biochemist in his or her final year before graduation. To avoid confusion, the term "pupil" is used to describe high school students.

2.2 SCIENTIFIC BACKGROUND

We will start this brief background survey by looking at the importance of plant lipids, from which many animal lipids are ultimately derived. Plant lipids are at the core of our existence, and without them we could not survive. Globally, plant lipids are the second most important source of edible calories in the human diet (after carbohydrates). Plant lipids are also sources of several essential vitamins and nutrients. For example, plant lipids are the ultimate source of the so-called "essential fatty acids" that are an obligatory component of the diet of all mammals—ever since that time many millions of years ago when our distant animal ancestors lost the ability to introduce double bonds beyond the Δ_9 position in long chain fatty acids. Since the dawn of agriculture, plants have been cultivated specifically for their lipid composition. The earliest olive plantations have been dated to more than 9,000 years before the present day, and the oilseed crop maize may have been domesticated in Mesoamerica as early as 10,000 years ago (Murphy 2007). In addition to their acyl lipid ingredients, plants are also important dietary sources of a host of other lipophilic compounds, including vitamins A and E and a range of phytosterols.

Most of our dietary plant lipid is derived from oil crops and is in the form of either "visible" (e.g., oils, margarines, chocolate) or "invisible" (cakes, cookies, confectionary, processed foods) fats. In the past, the lipid compositional requirements for these products have been provided by commodity plant oils that are blended or chemically modified (e.g., by hydrogenation) for a particular edible application. More recently, there has been a move toward a greater segmentation of the commodity plant oils market, with an increasing focus on the perceived health benefits of dietary lipids, especially from plants. Hence, we have an ever-increasing demand for plant oils that are enriched in monounsaturates, very long chain ω-3 or ω-6 fatty acids, carotenoids, phytosterols and tocotrienols. With an increased willingness by buyers to pay a premium for such nutritionally enhanced oils, it is becoming more economic for growers and processors to segregate such value-added products. This, in turn, is driving researchers and plant breeders to develop new varieties of oil crop designed for consumers who are becoming ever more aware of lipid-related nutritional issues, such as the presence of *trans* fatty acids and saturates in foodstuffs of all kinds.

Another interesting aspect of lipid science also related to nutrition is the question of the overall balance of fat in the diet and the chemical composition of such dietary fat. The past few years have seen an explosion of interest in the consequences of the so-called "obesity epidemic" that is especially acute in richer countries, with the United States leading the way in the incidence of life-threatening obesity in both adults and, more worryingly, children (Williamson, 2004). Colleagues at my own university have reported on the steep rise in type-2 diabetes in the local community, especially among poorer socio-economic groups (Davis et al., 2004; Siegel et al., 2005). They have also been involved in the development of innovative food products, such as low-fat cookies, and in outreach initiatives in the area of extending physical activity as a lifestyle-enhancing corollary to healthy eating (Brindley, 2004; Boobier et al., 2005, 2006).

Another research initiative has uncovered disturbing evidence of steroid abuse in the local community that probably reflects national and even international trends (Grace et al., 2003). This abuse started in fitness clubs as (mainly) men sought to improve their physical performance in athletic events, but has now extended to teenagers seeking cosmetic enhancement to their physique for social reasons (BBC, 2006). The side effects of steroid abuse can include heart disease, impotence, male lactation and female facial hair growth (Graham et al., 2006). As discussed below, sterols are key components of our diet and, of course, are also lipids. As well as being lipid-related, each of these issues, from obesity to steroid abuse, are immensely topical, are the subject of great public concern and are being actively researched by colleagues. As a lipid researcher, I find it both interesting and rewarding to engage in such research with colleagues in different, but related, fields. The creative use of such research synergies is a useful way of linking a public outreach program with taught courses and ongoing university research activities to ensure that our work is seen as interesting, relevant and important in the community, whether local, national or international.

2.3 KEY DISCUSSION TOPICS

In this section, I will introduce several lipid-related themes that we have used in a wider teaching/outreach context to relate lipid science to broader issues of public interest to our various audiences.

2.3.1 THE IMPORTANCE OF NON-ACYL LIPIDS

Non-acyl lipids are a useful topic to use in outreach programs because they impact on so many issues of concern, from the used of steroids in athletics doping to their dietary importance and supplementation in a new generation of genetically modified (GM) crops. Whereas in the past a lot of attention has been paid by researchers to modification of the relatively abundant acyl lipids in crops, there is now increasing awareness of the importance of many non-acyl lipids that may be present in insufficient quantities or may be removed or destroyed during the processing of plants for food. We will consider three examples here. First, there is the case of conventional rice, which is almost entirely deficient in carotenoids, such as β-carotene. This can result in vitamin A deficiency in those (mostly poor) people who are excessively reliant on rice in their diet. Second, vitamin E lipids are present in oil palm and can be manipulated to higher levels to serve as antioxidant-rich nutraceuticals. Third, several plant sterols and stanols that have well-characterized hypocholesterolemic functions are already produced in some plant tissues but may be engineered into a wider range of transgenic oil crops in the future (Broun et al., 1999; Corbin et al., 2001).

2.3.1.1 Golden Rice (Vitamin A)

Probably one the best-known recent examples of a nutritionally enhanced crop developed by genetic engineering is the transgenic "golden rice," which was produced by a Swiss-based group (Ye et al., 2000). This transgenic rice contains three inserted genes encoding the enzymes responsible for conversion of geranyl geranyl diphosphate to β-carotene. It is claimed that consumption of this rice by at-risk populations may alleviate the vitamin A deficiency (leading to night blindness) that currently afflicts some 124 million children worldwide. Interestingly, the rights for the commercial exploitation of golden rice in developed countries, including the USA and Europe, have now been acquired by the biotech firm Syngenta. This could possibly lead to the marketing of new types of "vitamin-enhanced" food products derived from golden rice, e.g., in breakfast cereals. Such products might be more acceptable to the public than the current generation of genetically modified (GM) foods, which are derived from crops that have only been transgenically modified for agronomic traits such as herbicide tolerance or insect resistance and therefore have no additional nutritional value compared with their non-GM counterparts.

One of the reservations expressed about the original varieties of golden rice was the relatively low content of provitamin A in the first versions of the GM rice plants. This problem has recently been solved by replacing a daffodil phytoene synthase gene with a similar gene from maize. The addition of this maize transgene to rice led to a 23-fold increase in provitamin A levels (Paine et al., 2005). It must be emphasized that this improved variety of golden rice still has many years of backcrossing into local varieties and field-testing before it will be known whether it is a viable and useful crop. One of the major challenges is to ensure that the provitamin A in the rice grains is in a form that can withstand processing, storage and cooking but is also completely bioavailable following consumption by people. This question of "bioavailability" is an important, and often overlooked, topic that can be used to generate debate with students. For example, many plants contain vitamins and mineral nutrients that are either lost during post-harvest treatments, such as processing and cooking, or pass through the human digestive system without being absorbed, e.g., due to chelation or other forms of chemical complexing. Probably the best-known example of this is spinach, which is touted as a source of dietary iron, but where only 2% of the iron in the leaves is actually bioavailable because most of it is strongly bound to oxalates and passes right through our digestive system without being absorbed. This means that a real-life Popeye would actually get very little strength-giving iron from his famous can of spinach. Interestingly, the more important nutritional benefit of spinach comes not from iron, but from its ω-3 fatty acid and fiber content (Murphy, 2004).

While transgenic golden rice is a promising development, there are other ways in which plant breeding can enhance vitamin A levels in food crops without recourse to the expensive and time-consuming transgenic approach, with all of its many complications regarding regulation, intellectual-property rights (IPR) and public perception. Several such initiatives are now under way and, although they have not attracted the massive media attention (both good and bad) that has been lavished on golden rice, some of these conventionally bred crops are much closer to release to farmers, most particularly in sub-Saharan Africa. For example, there are two programs aimed at enhancing vitamin A levels in the staple food crops maize and sweet potato. The first program is the High β-carotene maize (HBCM) Initiative, organized by the HarvestPlus consortium. The HarvestPlus consortium includes the United States Agency for International Development (US-AID); the International Food Policy Research Institute, Washington; and the International Center for Tropical Agriculture, Colombia (ICTA, which is part of the international network of Consultative Group on International Agricultural Research [CGIAR] centers). While one aim of the HBCM Initiative is to develop vitamin A-enriched varieties of maize, an equally important goal for researchers at Iowa State University is to assess the bioavailability of the vitamin A in the kinds of maize products actually eaten by the target population in Africa.

The second program, called Vitamin A for Africa (VITAA), is jointly funded by the Macronutrient Initiative and US-AID. As part of VITAA, the dietary impact in school-age children who were given high-carotene varieties of sweet potato was evaluated with promising results. Sweet potato is the fifth most important crop on a fresh weight basis and is especially important in Africa, where it is traditionally cultivated by women. Plant breeders from the International Potato Centre (CIP, which is another CGIAR centre) at Lima, Peru have recently produced new varieties of orange-fleshed sweet potatoes. The new sweet potatoes are highly enriched in a readily bioavailable form of vitamin A. These new clonal varieties are being mass propagated in Peru and sent for a 12-month quarantine period to Kenya, before being distributed to centers in Ethiopia, South Africa, Uganda and Tanzania for further evaluation and on-farm testing. The final stage will be to provide the sweet potatoes for cultivation by needy farmers throughout sub-Saharan Africa. VITAA researchers from Michigan State University and the CIP have concluded that the new orange-fleshed varieties could replace white sweet potato varieties, and thereby directly benefit over 50 million children under 6 years old currently at high risk from vitamin A deficiency (Low et al., 2001).

2.3.1.2 The Vitamin E Group

The vitamin E group of compounds includes four tocopherols and four tocotrienols, all of which have significant antioxidant properties. These lipophilic vitamins are implicated in reduction of blood cholesterol levels, plus protection against ageing and several forms of cancer. Vitamin E group compounds are found in most non-processed (i.e., cold-pressed or virgin) vegetable oils, but are often lacking in the vast range of prepared foods made from the more common, and cheaper, processed oils. There is interest in trying to increase the levels of this group of lipidic vitamins in plant oils using a variety of approaches. For example, transgenic plants that accumulate 10–15-fold higher levels of vitamin E compounds have been engineered by adding homogentisic acid geranylgeranyl transferase genes from several cereals to *Arabidopsis* plants (Cahoon et al., 2003). As with vitamin A, however, there are alternatives to the GM route that are worth exploring with students.

For example, unrefined palm oil contains significant amounts of vitamin E group compounds (Han et al., 2004). In recent years, breeders have identified several varieties of oil palm with oil that is highly enriched in tocols to levels in excess of 1500 ppm, which would be of great interest as potential health food products. The unprocessed, or "virgin," form of this oil is a visually striking burgundy-red color, due to the additional presence of nutritionally important antioxidant pigments including yellow/orange carotenoids and deep-red lycopenes. Lycopenes are the red pigments that are so prominent in tomatoes, cayenne and bell peppers, red grapefruit and a few vegetable oils, most especially from the oil palm. Although both carotenoids and lycopenes are available as vitamin supplements, by far the most reliable way to ensure their efficient uptake in the body is by consuming them as part of an original food product, such as fruits or unprocessed plant oils, rather than in isolated capsule form. This nutritious oil is already used by people in tropical Africa, but has not yet been taken up widely elsewhere, possibly because of an assumption by retailers that Western consumers already accustomed to the more common bland (in taste and color) light-yellow oils might be put off by the unusually vivid color.

These issues are good to discuss with students—for example, would you buy a bottle of red cooking oil or would you prefer to get your vitamins from a capsule? The question of red palm oil can also bring up developing-country topics because if such oils caught on with Western consumers it would greatly benefit poor farmers in producer countries in West Central Africa such as Sierra Leone, Liberia, Ivory Coast, Congo and Angola.

2.3.1.3 Phytosterols and Stanols

Another category of plant lipid of interest to the food industry is the phytosterols. Margarines enriched in phytosterols extracted from (non-transgenic) wood pulp or vegetable oils have recently been marketed and, despite an appreciable price premium compared with conventional margarines, they have enjoyed a good commercial success. The appeal of the phytosterol-enriched margarines is based on evidence that they may help to reduce blood cholesterol levels and hence combat heart disease. Such products could be made more cheaply if more of the phytosterols were synthesized in the same seeds as the oil from which the margarine is derived. Efforts are now under way to upregulate phytosterol biosynthetic pathways in transgenic plants. The impact on human health of such products could be considerable. Indeed, it has been surmised that the widespread availability and consumption of low-cost, phytosterol-enriched margarines could eventually lead to a quantifiable reduction in national rates of cardiovascular disease, which is still the most common cause of mortality, especially in low-income groups, in all industrial societies (Plat and Mensink, 2001).

2.3.2 Marine Lipids and Omega-3 Fats

Consumption of fish is currently recommended in most Western countries as part of a balanced diet, and much of the nutritional benefit of the fish actually comes from the very long chain polyunsaturated fatty acids (VLCPUFAs) of the fish oils. Oils rich in omega-3 (ω-3) fatty acids include the so-called "fish oils" (or more correctly "marine oils"), which are characterized by relatively high levels of VLCPUFAs such as eicosapentaenoic acid (20:5ω-3, EPA) and docosahexaenoic acid (22:6ω-3, DHA). These compounds are part of the group of ω-3 fatty acids that are essential components of mammalian cell membranes, as well as being precursors of the biologically active eicosanoids and docosanoids (Funk, 2001; Hong et al., 2003). There have been numerous reports concerning the importance of dietary supplementation with these fatty acids for human health and well-being. For example, dietary VLCPUFAs have been shown to confer protection against common chronic diseases such as cardiovascular disease, metabolic syndrome and inflammatory disorders, as well as enhancing the performance of the eyes, brain and nervous system (Crawford et al., 1997; Benatti et al., 2004; Spector, 1999).

These fatty acids can be synthesized by the fish themselves or derived from microorganisms, especially photosynthetic microalgae, that are ingested as part of their diet. As an alternative to fish consumption, therefore, it is possible to purchase VLCPUFA dietary supplements that are derived from cultured microalgae or fungi. However, low oil yields and high costs of oil extraction have limited the scope for this production method, and ever-dwindling fish stocks are also threatening supplies of the main source of marine oils. This situation has led to renewed interest in the possibility of breeding oilseed crops that are capable of producing significant quantities of VLCPUFAs in their storage oils. Higher plants do not normally accumulate such fatty acids, but can accumulate C22 and C24 monounsaturates and C18 polyunsaturates in their seed oils, so it seems possible that C20 and C22 polyunsaturates might also be accumulated, providing the plants were able to synthesize these fatty acids.

It is worth mentioning here that none of these VLCPUFAs are strictly "essential" in the diet in the same way that true vitamins are. The only unequivocally essential fatty acids are linoleic acid (*cis* 9, 12 18:2) and α-linolenic acid (*cis* 9, 12, 15) ω-3 and ω-6 fatty acids, respectively, which mammals are unable to synthesize. These essential fatty acids are common in grains and leafy vegetables. In a well-nourished and healthy individual, all of the VLCPUFAs can be potentially synthesized from linoleic acid and α-linolenic acid. Unfortunately, many Western diets do not deliver a balanced spectrum of fatty acid intake, so VLCPUFAs and other fatty acid supplements are sometimes required to maintain optimum health. An example of another fatty acid supplement is gamma (γ)-linolenic acid, or GLA, an ω-6 fatty acid that is usually obtained from evening primrose or borage oil. GLA supplements are taken for almost every ailment under the sun, but a healthy person with a balanced diet should be able to synthesize enough GLA for all normal metabolic functions without recourse to such nutritional supplements. The cases of GLA and the other supposedly "essential" fatty acids can be an interesting topic for classroom discussion.

2.3.3 Low *Trans* Fats

The market for oils that contain reduced or zero levels of *trans* fatty acids is driven by health concerns that have led to the imposition of labeling requirements to state whether a product contains over a given threshold of these fatty acids. Such labeling requirements were introduced into the USA in January 2006 and are likely to be required in the European Union in the near future. Typical threshold levels of *trans* fatty acids that would trigger compulsory labeling are in the region of 0.5 to 1.0%, whereas some existing foods can contain as much as 40% *trans* fatty acids. The solution in most cases will be to develop high-oleic oil crops and, as we have seen above, breeders have been gradually producing such varieties of the major oilseeds over the past decades. There are still challenges for breeders to reduce further or to eliminate altogether α-linolenate from seed oils and to ensure that the high-oleic traits are crossed into their highest-performing elite commercial lines. One of the most interesting aspects of the *trans* fatty acids issue is that researchers have known about potential problems with these fats for several decades but they have only recently emerged into public consciousness and hence become an item of concern to consumers.

It would be interesting to discuss with students why people have apparently been allowed to eat foods enriched in *trans* fatty acids for over a century with little or no health concern. As with the fish oil issue, there is also the question of proportion. Is it better to consume fish oils and risk ingesting chemical residues or to go without and risk dietary deficiencies? Also, is it better to consume *trans* fatty acids in a high-polyunsaturate margarine or use butter instead, with its high levels of saturates and cholesterol? One interesting aspect for discussion is whether it is enough to simply label foods containing *trans* fats, as is the case in the USA, or whether governments should be more proactive in issuing dietary advice, as in Canada and parts of Europe. Also, to what extent should governments be allowed to enforce healthy eating, for example by not merely labeling but actually banning some "unhealthy" foods, such as those with high saturate or *trans* fatty acid contents? The case of Denmark and its 2003 ban on *trans* fats can be used in such a discussion.

Some of the better Internet resources in this area include UK Food Standards Agency http://www.food.gov.uk/news/newsarchive/2004/jun/fishportionslifestagechartandhttp://www.food.gov.uk/news/newsarchive/2004/jun/oilyfishwebcast; tfX: the campaign against trans fats in food, http://www.tfx.org.uk/page116.html; Fats of Life http://www.fatsoflife.com/; and Health Canada http://www.hc-sc.gc.ca/fn-an/nutrition/gras-trans-fats/index_e.html.

2.4 CASE STUDY OF AN OUTREACH ACTIVITY: GENETIC ENGINEERING AND LIPIDS

2.4.1 BACKGROUND

The teaching of molecular biology, biochemistry and biotechnology at high school and undergraduate levels can sometimes be challenging to educators. These topics are often regarded as technically demanding by students, and many courses and textbooks dwell on the molecular mechanisms underlying biological development, most especially focusing on gene function and metabolic pathways. However, at the same time, many students who may not be "natural" molecular biologists are intensely interested in the wider issues raised by, for example, human cloning or GM crops. Biotechnology and biochemistry are both, therefore, relatively difficult technical disciplines, but also ones that raise many complex and contentious questions in areas as diverse as public awareness, commerce, ethics and politics (Polkinghorne, 2000). Previous experience of teaching biotechnology and related subjects gave us anecdotal evidence that a broadening of teaching methods to encompass direct discussion and role-playing can improve student-learning outcomes. This was true for students in general, but was especially so for those who were less comfortable with the conventional didactic approach to the subject.

To study further the utility of broader teaching methods in these subjects, we initiated a university–high schools partnership in biotechnology. We studied the impact on learning experiences of third- (final-) year biotechnology undergraduates when they were given direct practical experience of explaining and discussing biotechnology issues, especially lipid-related topics, to high school pupils. We hypothesized that the need to develop teaching resources and to communicate with high school pupils might improve overall student comprehension, not only of more general issues, but also of the core scientific content of molecular biology and biotechnology. A major aim of our study was to investigate the impact on the development of additional cognitive and communication skills in third-year Biology Honors undergraduates by challenging them to discuss issues relating to GM crops with high school pupils. As a further challenge to the students, we specifically selected schools from socially disadvantaged areas of the Welsh Valleys having historically low participation rates in further and higher education (Further Education Funding

Council for Wales, 2000). Our major engagement tools were on-campus Masterclasses and School Visits.

2.4.2 The Study: Masterclasses and School Visits

Masterclasses were structured, daylong sessions in which groups from several schools, numbering up to 80 pupils and teachers, visited the university campus. Each Masterclass included a seminar, visits to research and teaching labs, and a full-scale debate on the motion "biotechnology has gone too far." A typical debate format involved four speakers, two speaking for the motion and two against. Following the speakers' presentations, the debate would be thrown open to discussion from the audience. Students role-played as GM proponents or opponents and some debates were livened up further by inviting anti-GM activists or spokespersons for GM companies to speak either for or against the motion.

School Visits were undertaken by third-year students, either individually or in pairs, to selected schools in the local area, i.e., within a 30-minute traveling time. Pairing of less confident students with more confident peers was a useful way to ensure that all students participated in the program. Visits were normally of about 2–3 hours and for each class they included a 20-minute presentation, after which the class was invited to complete questionnaires and engage in discussion on the issues raised. The pupils could ask questions on the presentation, testing the student's background knowledge and communication skills. Discussions were often initiated by the teachers, who had also listened to the presentation, thereby ensuring a productive visit for the student. These activities were designed to cover a single lesson period, allowing for the involvement of up to three classes during one school visit. The arrangement of these visits required a degree of organization from the students and liaising with the teachers to ascertain suitable times, lesson structure and class sizes. Students helped to develop questionnaires, information leaflets and a PowerPoint presentation that were then used in schools to support the interactive sessions with the pupils. A useful background text for students to use in the formulation of their discussion session was Barnes and Todd (1977) supplemented by the more recent reports by Simonneaux (2002a, b).

2.4.3 Resources

The administrative infrastructure for our outreach program, of which these activities are only one example, was provided by the university Centre for Lifelong Learning (CELL), which arranged Masterclasses and room bookings, etc. A useful resource that was generated and continually refined during the program is a brochure for teachers and students. Called "Genetic Engineering: Monster or Marvel," this short color brochure explores the background to key GM-related issues and can be customized for specific activities. To cut down on costs, the brochure is printed by our university Reprographics Unit and its design and contents can be a useful topic of discussion with pupils, teachers and students. For example, in one assignment, third-year students were challenged to produce their own design for brochures that explored the issue of golden rice and vitamin A in developing countries.

2.4.4 Monitoring by Questionnaires

The results of Masterclasses and School Visits on pupil perception of GM-related issues were monitored by a series of questionnaires supplemented by the recording of direct feedback from pupils. In most cases, three questionnaires were used for each group of pupils. The first questionnaire was sent to schools for completion about a week before pupils were exposed to any new teaching materials. Pupils were then sent brochures designed to supplement their preparation work with teachers. Pupils completed the second questionnaire on the day of the activity, immediately after its completion. This questionnaire was designed to assess the effects of the activity on pupils' perceptions of GM-related issues. We were particularly interested in any topics where pupils had changed their opinions as a result of their interaction with the student. The third questionnaire was completed by pupils several months after the activities and was designed to measure the extent to which changes in pupil perception may have altered over the longer term, rather than immediately after the activity, i.e., were their opinions merely altered transiently by the activity or are any observed changes more long lasting?

Topics covered in these activities included all the examples given in previous sections, such as GM crops with high vitamin A (golden rice), high VLCPUFA (fish oils) and high stearic acid (low *trans*), as well as non-lipid topics such as vaccines and biodegradable plastics from GM crops. Even in the latter cases, however, links could be made with lipid science. For example, in GM crops, biodegradable plastics such as polyhydroxyalkanoates are deposited in seeds instead of storage lipids and early attempts to achieve this were failures due to a lack of understanding of the basic metabolism of acetyl-CoA to fatty acids or flavonoids (Rezzonico et al., 2002). Students wrote up their assignments in the form of an analysis of the questionnaire data and a "reflective essay" on the whole experience and how it impacted on their comprehension of and overall interest in biotechnology and biochemistry. As shown below, there were many positive comments on the framing of their work within more generalized social and ethical contexts, and how this improved their own interest in "hard science" topics such as lipid biochemistry and biotechnology. Such views were most common in students who were weaker in conventional testing situations, showing that they might be better engaged by broadening the scientific learning experience to include such wider issues.

2.4.5 Examples of Student Data Obtained from Masterclasses and School Visits

Comments made by participating pupils:

- "I found the debates more useful than the talks and would like to have more in the future."
- "I found the debates more interesting and would like to have more talks in the future covering topics not covered at school such as bioplastics."
- "I think the Masterclasses are a good idea."

Feedback from participating teachers:

- Teachers used both activities as supplementary lessons to support their formal science curriculum teaching.
- They commented that the students had pitched their presentations at the right level for pupils to understand.
- They wanted to be involved in any future activities.

2.4.5.1 Analysis of the Questionnaires

In general, pupils thought that both the Masterclasses and School Visits helped them better understand the issues covered, and that these activities were useful. For example, after the Masterclasses, 88% of pupils reported that they understood more about biotechnology and 86% said they were better informed. On the wider issues discussed, pupils said that GM foods could be beneficial to developing countries. However, they did not believe that GM food could "feed the world." This suggests that they were aware of and were able to discount some of the more exaggerated claims made on behalf of GM foods. The pupils thought that any reservations they held about the technology would be unlikely to influence government policy in this area. From the Masterclasses, 81% of pupils thought that GM was an important issue to discuss but only 29% of them thought that their views could have an effect on whether GM foods and crops are eventually accepted in the UK.

The data generated from questionnaires completed before and after each Masterclass were compared in order to assess its immediate impact on pupil opinion. A particularly important finding was that the proportion of pupils who were undecided in many of the key topic areas dropped considerably after the events. In many cases, the pupils became more enthusiastic about some of the potential benefits of GM technology immediately after the Masterclass. However, not all opinions changed in the pro-GM direction. Pupils tended to become more cautious when asked if they would eat GM foods—whereas 46% said they would eat GM foods before the Masterclass, only 36% would do so afterward. Interestingly, the general tendency to be more supportive of GM technology immediately after the Masterclasses was not maintained in the first follow-up surveys of pupils conducted 6 months after one of the Masterclasses (during March 2003). Here, pupils became increasingly undecided on the topics that they had been initially enthusiastic about after attending the Masterclasses. For three of these questions, pupils tended toward being more "undecided" after the follow-up study. We interpret this as a positive change of opinion to a less "black-and-white" pro- or anti-GM stance (perhaps after recognizing the complexities of the issues) rather than as a mere abstention from any opinion at all.

Comparison of questionnaire data before and after the School Visits demonstrates that these activities have helped the pupils form opinions, as reflected in the lower numbers of pupils' being "undecided" after the events. The overall trend in opinion change for this study was in favor of GM technology. As with the Masterclasses, a 3-month follow-up study showed increasing uncertainty in some areas, e.g., about whether GM foods could be more nutritious and therefore healthier. However, at the same time, pupils became significantly more supportive of statements like "some types of GM food may be good" (increased from 77% to 100%) and "I approve of GM on practical and scientific grounds" (increased from 39% to 75%).

2.4.6 Conclusions from Student/Pupil Program

- Course marks and general motivation of participating undergraduates were significantly improved compared with previous performance.
- Both the Masterclass and School Visits formats were quantitatively effective in terms of pupil outcomes as measured by analysis of questionnaires.
- The School Visits format is more flexible, is more suitable as an individual student activity and is more easily assessed as an independent activity.
- Both activities were strongly supported by teachers, who also wished to participate in similar schemes in the future.

The pilot study described here shows that relatively open-ended activities involving outside contacts with high schools can be successfully incorporated into a full 20-credit, third-year biotechnology module. At first sight, it may appear that the additional workload involved in a Schools Visit program, including all the associated preparation and analysis, may divert students from the primary scientific goals of the module. However, we found that this was not the case: indeed, student enthusiasm for and comprehension of the "hard science" parts of the module were definitely improved by their participation in the school-related activities. The students needed to master enough science to be able to design their presentations and to engage in informed discussions with pupils. This stimulus gave them a strong motivation, not only to understand the lectures, but also to read around the topic and especially to research into public concerns and ethical implications of biotechnology. An alternative format to teaching wider principles of biotechnology transfer is the service-learning approach as recently described by Montgomery (2003). This can involve community-based activities using partners such as local non-profit organizations rather than schools. Therefore, the use of any one of a variety of approaches can result in favorable outcomes for students. In our case, we preferred a school-centered approach as a way of addressing additional issues in relation to low rates of educational achievement in our local region.

The data analysis used in this study involved skills that were also used in other course modules, e.g., ecology. However, by using these techniques to analyze pupil perceptions, the students were strongly reminded of the limitations and potential misinterpretations involved in such activities as opinion polls and consumer surveys. Students also reported personal satisfaction at being able to engage in constructive discussions with teachers and pupils in an area where, for once, they were regarded as "the expert." It is not possible to quantify the extent of empowerment felt by students but it is apparent that participation in such activities can bolster their confidence and general motivation. We are also aware that the students gain useful transferable skills in a wide variety of areas from public speaking to preparation of handouts and brochures.

Another attraction of these sorts of activities is that, in addition to being beneficial to our own undergraduates, they also succeeded in stimulating interest in biology among high school pupils as well as involving their teachers. Our first partner schools were in the relatively deprived former mining communities in the Welsh Valleys within 10 miles of the university. Rates of entry into tertiary education (especially universities) from such areas are well below the national average and are the subject of great concern, both in Wales itself (in the National Assembly) and in the UK as a whole (Further Education Funding Council for Wales, 2000). Despite these gloomy statistics, we found both pupils and teachers to be receptive and keen to engage in dialog with our students about biotech and lipid-related issues—provided they were "packaged" appropriately. In other words, presented not just as scientific topics, but rather in their wider social context that could include medical, ethical and economic dimensions.

2.5 OVERALL CONCLUSIONS

When I was studying biochemistry as an undergraduate, lipid science was almost universally regarded as one of the most difficult and tedious courses. Luckily, I was taught by several inspirational professors who introduced the new and exciting concepts of the fluid mosaic model of membrane organization and the dynamic role of lipids in nutrition. Thanks to this experience I subsequently embarked on a career of research and teaching that was very much focused on lipids, and especially plant lipids. In the same way, as educators and communicators reaching out to ever-wider audiences, we should seek to tell many of the fascinating stories that lipid research has recently uncovered. As discussed in this and other chapters of this book, lipid-related topics are matters of immense public concern that are mentioned in some form in the media on an almost daily basis.

Moreover, although they might not know it, many of our colleagues also work in lipid-related areas. Over the past few years I have worked with epidemiologists, care scientists, nutritionists, clinicians, exercise specialists and many others on lipid-related topics, some of which are described above. It is important that, in seeking to reach out to the wider community to make their subject more accessible, lipid scientists also recruit such colleagues in related academic disciplines. My own experience is that once people realize that a topic such as lipids is relevant to their daily lives, they are much more likely to seek to learn more about it, not because it is part of any curriculum, but rather to satisfy their own interest and curiosity. Such learning experiences also tend to be more enduring and enjoyable than the sort of pre-exam cramming that is still all too prevalent in universities and high schools.

ACKNOWLEDGMENTS

We thank staff at the University of Glamorgan Centre for Education and Lifelong Learning for their assistance with the Masterclasses. Financial support outreach program is from the Welsh Assembly Government, Wales Science Year, Biochemical Society, British Association, and the Learning and Teaching Support Network (LTSN).

REFERENCES

Note: The account of this case study is based in part on our previous article: Todd A and Murphy DJ (2003) Evaluating University Masterclasses and School Visits as mechanisms for enhancing teaching and learning experiences for undergraduates and school pupils. *Biosci Educ Electron J*, published online, November 2003, http://bio.ltsn.ac.uk/journal/vol2/index.htm. Accessed 25 September 2006.

Baker JS, Graham M and Davies B (2006) *J Royal Soc Med* 99, 330–331, www.rsm.ac.uk/media/downloads/j06-07gym.pdf. Accessed 25 September 2006.

Barnes D and Todd F (1977) *Communication and Learning in Small Groups*, Routledge, London.

BBC (2006) Bodybuilders "breast drug abuse," BBC news online June 14 2006, http://news.bbc.co.uk/1/hi/health/5080640.stm. Accessed 25 September 2006.

Benatti P, Peluso G, Nicolai R and Calvani M (2004) Polyunsaturated fatty acids: Biochemical, nutritional and epigenetic properties, *J Am Coll Nutr* 23, 281–302.

Boobier WJ, Baker JS and Davies B (2005) Obesity and related medical conditions: A role for functional foods, *Publ Hlth Nutr* 8, 1328–9.

Boobier WJ, Baker JS and Davies B (2006) Development of a healthy biscuit: An alternative approach to biscuit manufacture, *Nutr J* 5, 1–7, http://www.nutritionj.com/content/5/1/7. Accessed 25 September 2006.

Brindley M (2004) "Oh yes! Welsh expert invents fat-free cakes," *Western Mail*, October 7 2004, http://icwales.icnetwork.co.uk/0100news/0200wales/tm_objectid=14726876&method=full&siteid=50082&headline=oh-yes--welsh-expert-invents-fat-free-cakes-name_page.html. Accessed 25 September 2006.

Broun P, Gettner S and Somerville C (1999) Genetic engineering of plant lipids, *Ann Rev Nutr* 19, 197–216.

Cahoon EB, Hall SE, Ripp KG, Ganzke TS, Hitz WD and Coughlan SJ (2003) Metabolic redesign of vitamin E biosynthesis in plants for tocotrienol production and increased antioxidant content, *Natur Biotechnol* 21, 1082–1087.

Corbin DR, Grebenok RJ, Ohnmeiss TE, Greenplate JT, Purcell JP (2001) Expression and chloroplast targeting of cholesterol oxidase in transgenic tobacco plants, *Plant Physiol* 126, 1116–1128.

Crawford MA, Costeloe K, Ghebremeskel K, Phylactos A, Skirvin L and Stacey F (1997) Are deficits of arachidonic and docosahexaenoic acids responsible for the neural and vascular complications of preterm babies? *Am J Clin Nutr* 66, 1032S–1041S.

Davis RE, Morrissey M and Currie C (2004) The epidemiology of hypoglycaemia in type 2 diabetes. Oral poster at the European Association for the Study of Diabetes Conference (EASD) Munich, Germany, September 5–9th 2004.

Funk CD (2001) Prostaglandins and leukotrienes: Advances in eicosanoid biology. *Science* 294, 1871–1875.

Further Education Funding Council for Wales (2000) Further and higher education statistics in Wales: 1998–1999, http://www.hefcw.ac.uk/FinanceAssurance_Docs/HE_FE_training_statistics_wales_1998_99.pdf. Accessed 25 September 2006.

Grace F, Baker JS and Davies B (2003) Evidence of hyperhomocystemia following long term anabolic androgenic steroid (AAS) use. Paper presented at the 9th European nutrition conference Rome, Italy, October 1–4. *Ann Nutr Metab* 47, PS. Q 8 pp 609.

Graham MR, Grace FM, Boobier W, Hullin D, Kicman A, Cowan D, Davies B and Baker JS (2006) Homocysteine induced cardiovascular events: A consequence of long term anabolic-androgenic steroid (AAS) abuse, *Br J Sports Med* 40, 644–648.

Han NM, May CY, Ngan MA, Hock CC and Ali Hashim M (2004) Isolation of palm tocols using supercritical fluid chromatography, *J Chromatogr Sci.* 42, 536–539.

Hong S, Gronert K, Devchand PR, Moussignac RL and Serhan CN (2003) Novel docosatrienes and 17S-resolvins generated from docosahexaenoic acid in murine brain, human blood, and glial cells. Autacoids in anti-inflammation, *J Biol Chem* 278, 14677–14687.

Low J, Walker T and Hijmans R (2001) The potential impact of orange-fleshed sweet potatoes on vitamin A intake in sub-Saharan Africa. Paper presented at a regional workshop on food-based approaches to human nutritional deficiencies, *The VITAA Project,* 9–11 May, Nairobi, Kenya.

Montgomery LB (2003) Teaching the principles of biotechnology transfer: A service approach, *Electron J Biotechnol* 6 (1), 13–15. http://www.ejbiotechnology.info/content/vol6/issue1/teaching/01. Accessed 25 September 2006.

Murphy DJ (2004) Overview of applications of plant biotechnology, in *Handbook of Plant Biotechnology*, eds. Christou P and Klee H, Wiley, New York.

Murphy DJ (2005) *Plant Lipids: Biology, Utilisation and Manipulation*, Blackwell, Oxford.

Murphy DJ (2007) *People, Plants, and Genes: The Story of Crops and Humanity*, Oxford University Press, Oxford, UK.

Paine JA, Shipton CA, Chaggar S, Howells RM, Kennedy MJ, Vernon G, Wright SY, Hinchliffe E, Adams JL, Silverstone AL and Drake R (2005) Improving the nutritional value of golden rice through increased pro-vitamin A content, *Nature Biotechnol* 23, 482–487.

Plat J and Mensink RP (2001) Effects of plant sterols and stanols on lipid metabolism and cardiovascular risk, *Nutr Metab Cardiovasc Dis* 11, 31–40.

Polkinghorne JC (2000) Ethical issues in biotechnology, *Trends in Biotechnology* 18, 8–10.

Rezzonico E, Moire L and Poirier Y (2002) Polymers of 3-hydroxyacids in plants, *Phytochemistry Reviews* 1, 87–92.

Siegel SR, Thomas NE, Cooper SM, Reyes MP, Barahona EC, Williams SPR, Baker JS, Davies B, Malina RM (2005) Prevalence of overweight and obesity in Welsh and Mexican school youth. 52nd Annual Meeting of the American College of Sports Medicine, Gaylord Opryland Hotel, Nashville, Tennessee, *Med Sci Sports Exerc* 37:5 Suppl.

Simonneaux L (2002a) Analysis of debating strategies in classroom in the field of biotechnology, *J Biolog Educ* 37 (1), 9–12.

Simonneaux L (2002b) Analysis of didactic strategies to help pupils develop argumentation skills in the field of biotechnology. *"BioEd 2000"* The Challenge of the Next Century, Paris, May 2000, www.iubs.org/cbe/pdf/simonneaux.pdf. Accessed 22 September 2006.

Spector A (1999) Essentiality of fatty acids, *Lipids* 34, 1–3.

Williamson D (2004) "Academic attacks attitude to obesity," *Western Mail*, May 28, 2004, http://icwales.icnetwork.co.uk/0100news/0200wales/tm_objectid=14282476&method=full&siteid=50082&headline=academic-attacks-flabby-attitude-to-new-killer-disease-of-obesity-name_page.html. Accessed 24 May 2007.

Ye X, Al-Babili S, Kloti A, Zhang J, Lucca P, Beyer P, Potrykus I (2000) Engineering the provitamin A (β-carotene) biosynthetic pathway into (carotenoid-free) rice endosperm. *Science* 287, 303–305.

3 Mentorships and Related Programs Provide Mechanisms for Involving Students in the Science of Fats and Oils

Norm Lee[*] *and James K. Daun*

CONTENTS

3.1 INTRODUCTION

The purpose of this chapter is to share a number of science- and technology-awareness programs in which the authors have been personally involved, including the programs of MindSet, the Manitoba Network for Science and Technology

* Corresponding author.

(www.mindset.mb.ca), to which both authors have made substantial contributions of their time and expertise. In some instances, these programs were tied directly to oilseed chemistry activities but many of the programs are more general in nature but eminently applicable to involving young people in careers in oilseeds chemistry.

Worldwide shortages of scientists, engineers and technologists are being predicted in many scientific disciplines by organizations such as United Nations Education, Scientific and Cultural Organization (UNESCO; http://portal.unesco.org/en/ev.php). At the same time, science and technology have become the drivers of economic and social development for the global economy. Countries not able to compete in these attractive areas of growth in science and technology are more likely to face a lower standard of living and quality of life than their economic rivals. The pressure to prepare more young people to follow careers in science and technology, both generally and specifically in the field of oil chemistry, comes from at least two sources—the need to replace the retiring cadre of scientists and the general need for more scientists as science becomes more of a driver of economic development. Countries that have relied on immigration of scientists may especially feel the pressure to produce more scientists as more nations are able to offer attractive and rewarding careers. More scientists will find staying at home in their own country as rewarding as working elsewhere, most likely in a more industrialized nation. There are, of course, other variables related specifically to oil chemistry and science. Despite the importance of fats and oils, and in particular oilseeds, to the economy of Canada, there has been little effort in developing programs to teach the chemistry and technology of these components, especially at the post-secondary level of education. Even basic education in lipid chemistry and biochemistry has been noted as deficient by Canada's Expert Committee on Fats, Oils and Other Lipids.[1]

Science knowledge is growing rapidly in many areas. The way science is being accomplished is also changing quickly. At the same time, schools and curriculum are notoriously slow to change and reflect these changes. For example, research in both Canada and the United states indicates that it takes between 16 and 23 years for new ideas to develop critical mass and become mainstream in education. The cumulative result of these trends is that students are more likely to be taught about the way science was, rather than the way it is and the way it is going to be.

MindSet (Manitoba Network for Science and Technology) has been held up as a model for creating programs to make young people more aware of the career and educational opportunities related to science and technology. While its programs cross many areas of science and technology including information and communications technology, interactive digital media, new materials and composites, advanced manufacturing and aerospace, it is most active in life science. Being located in Manitoba, this work includes agriculture and, within that, the oilseeds sector.

The situation is not likely to get better quickly in Canada—and many other places—as long as young people have negative opinions about training and careers in these areas. Negative opinions about science have their beginning in several places. Scientists are often portrayed in the media as "bad." In fact, one study of Canadians, published in *The Toronto Globe and Mail,* indicated that more than 60% of Canadians thought scientists were "evil." Befitting this attribute, science writer Jon Franklin noted that about 10% of scientists in movies and television are killed before the end of the show. He also noted sarcastically that this was "some career motivation."

Negative images of science are evident in the fats and oils area. Recent portrayals by the popular press of lipids such as *trans* fatty acids, saturated fats and cholesterol have ingrained a confused and negative image of this entire area in the minds of many consumers. Any lipid chemist who has ever attended a party along with people who are not familiar with the area, including chemists, medical professionals and biochemists, will find themselves spending time explaining the role of different fats in preventing high cholesterol. Even "good fats" can receive negative attention. Recently, canola oil, cited by the American Heart Association as a desirable vegetable oil, received considerable bad press through a malicious web site. The Canola Council of Canada spends considerable time providing scientific refutations to this unfortunate information, see http://www.canola-council.org/cooking_myths.html.

In some countries, the ability to attract, develop and retain high quality professionals is tied to weak science programs in the schools, especially in the elementary and middle schools. Even at the high school level, educators, including counselors, are not aware of the needs of industry and lack of awareness about career opportunities. Teachers are the "gatekeepers" to students, and if they pass on negative connotations about science, students are much less likely to pursue careers in that area.

3.2 AN INTEGRATED APPROACH

Some general things embodied in the work of MindSet seem related to its success, most importantly the development of partnerships across many sectors including schools, post-secondary, the business sector and government. Considering educators as part of the science and technology workforce is another aspect of a strategic plan to develop a stronger science and technology culture within its jurisdiction. Another variable, not often mentioned, is emphasizing the interrelatedness of science and technology, something that does not appear to be happening enough in schools. Today's scientists do what scientists have always done—observe, hypothesize, experiment, calculate, problem-solve and share. However, these efforts are reinforced by the use of technology. In some ways, no world-class research is being done without technology, in oilseeds chemistry or otherwise. This relationship is not being emphasized in schools. For that matter, the science teachers do not team up with the technology people to show students how technology contributes a great deal to modern-day science. Further, computers are used more for research than they are used as scientific instruments. Some science teachers don't use computer technology themselves.

Understanding globalization is not really a science concept. On the other hand, we all have to understand that the world is changing, getting "flatter," as Thomas Friedman of the *New York Times* describes it. Technology has made the economic playing field more level—flatter. In a very real sense, we compete with everyone in the world doing something similar to what we do. The same technology that has made the world immeasurably more competitive can also make the world immeasurably more cooperative.

One way of bringing home the concept of globalization to students is to involve them in international science projects. Some of these are competitive, such as the International CyberFair, for which students can highlight some aspect of their life space, e.g., local industry. Besides supporting these competitive programs, MindSet has worked with schools to get them involved in international partnerships with other schools, including projects with Israel, New Zealand and the United States. For example, the project with Israel compared the ability of different crops found in each country to produce bio-energy.

3.3 MENTORING AND COACHING

Common threads running through many of the programs described here are the provision of mentoring and coaching as important ways to get students really involved in science programs. For whatever financial or philosophical reasons, many schools do not provide enough authentic scientific experiences for students. In some instances, science is not considered important, so very little gets taught. Often, the teachers assigned to teach science do not have a scientific background and have no interest in it. Budgetary restraints may limit the purchase of scientific equipment so science is taught from a textbook with little hands-on experience. In these situations, school does not provide students with a "feel" for being involved in science or what it really means to be a scientist.

Mentorships are a great way to provide students with an opportunity to try a science career on for size. Students most often work in a lab part time alongside older, more accomplished people. They use scientific instruments and other forms of technology often not found in schools. During mentorships, students may get a feeling of having discovered something or the joy of completion. There are documented cases of students' doing research that leads to a patent for them. Mentorships can also lead to summer employment in the mentor's facility.

Another form of supporting student development is coaching them for various tasks that lead them toward a science career. Mentorships tend to be more long-term and general, while coaching tends to be more short-term and focused on a specific skill. There are many science activities open to students during the course of a school year. These range from long-term commitments such as science fair projects to shorter-term activities such as essay writing and various kinds of Internet activities. Usually, many students consider participating in these opportunities but, in the end, only a small number end up trying the activity. There are many reasons that more young people don't get involved, but not knowing what is expected and the fear of failure that goes with that are probably the most common.

MindSet has supported the involvement of many students in science activities. The coaching and support that is provided has resulted in a disproportionate number of students' achieving at high levels nationally in such contests as essay writing and web site contests. Student workshops have been offered in many areas, including preparing scientific posters, development of research skills, technical writing, writing for the Internet and improving creativity. While some of this coaching has been provided by MindSet itself, most often MindSet contacts people who have more expertise in a given area so that students have a chance to meet outstanding professionals.

3.3.1 Bio-Trek

Bio-Trek is a summer camp program that offers hands-on biotechnology experiences to high school students. Since the program was begun by Dr. Sheppy Coodin in Winnipeg in 1999 at St. John's Ravenscourt School, more than 500 students have been involved in other locales, including Toronto, Ottawa and Oakville in Ontario, Steinbach in Manitoba; Calgary and Edmonton in Alberta; Montreal, Quebec and Vancouver in British Columbia. As well as being exposed to careers in science, students were involved in activities such as purifying deoxyribonucleic acid (DNA) from cells, turning genes "off" and "on," and inserting foreign DNA into *Escherichia coli* bacteria.

3.3.2 University and College Outreach Programs

Many universities and colleges across Canada offer summer programs similar to Bio-Trek as part of their community outreach. Mini-universities and junior colleges provide hands-on science and engineering activities in many areas, including agriculture and chemistry. Most of the courses are given on campus, with university students delivering the activities created by professors and instructors.

3.3.3 Sanofi-Aventis Biotalent Challenge

The Sanofi-Aventis Biotalent Challenge (SABC) is one of Canada's most prestigious student science programs, attracting many of the best and brightest students from across the country. When 350 past participants of the program were surveyed 5 years after they had competed, the students' responses indicated the impressive results:

- 74% indicated that it helped plan studies or careers in fields such as biotechnology, health care, agriculture and the environment.
- 76% had received at least one scholarship, bursary or other academic achievement or recognition.
- 85% said SABC provided a positive understanding of Canada's biotech industry.

SABC, previously known as the Aventis Biotechnology Challenge and, previous to that, the Connaught Student Biotechnology Exhibition, was created in Canada to encourage senior high school students to participate in hands-on biotechnology research. Students are invited to submit their ideas for an experiment they would like to do, based on their independent research from journals, newspapers and other media as well as from the Internet or from people in their own lives such as teachers or parents. These experimental ideas are then evaluated by a blue ribbon panel of scientists who approve the feasibility of the research. Some experimental ideas are accepted immediately. Some are sent back for some reworking. Some are rejected outright as being impossible.

The approach has a unity to it. The whole process mirrors the steps that scientists go through in creating their own research, starting with developing an experimental question from research and submitting the idea to an organization to evaluate the idea for funding. Students receive about $200.00 to assist in the research if their suggestions are accepted. Further, whenever possible, the organizers of the program in each province find mentors for the students.

Most often, the mentor will be one who does research in the area of the student's work. Sometimes, it is not possible to find someone locally doing the exact type of work, so once in a while, a mentor is found in another city and the mentoring is carried out via e-mail or some other type of communication. Very rarely, the student ends up doing the work alone. On average, the student works under the mentor's guidance for 5 months before competing with other students in their area for the best work.

There are SABC programs in 13 cities in Canada, each hosted by a local organization interested in science. The winner of each region competes nationally for further prize money. The national competition is carried out via videoconferencing using the facilities of National Research Council Institutes in Canada. The national winner has the opportunity to go on to compete internationally at the Sanofi-Aventis International BioGENEius Challenge, organized by the Biotechnology Institute located in Arlington, Virginia on behalf of Sanofi-Aventis. Regional winners from Canada and the United States compete at the Biotechnology Industry Organization annual convention. For this third level, an outstanding group of scientists, educators and businesspeople involved in biotechnology judge the projects. The top five winning experiments get major cash prizes but all students have the opportunity to display their work to between 1,500 and 20,000 delegates who attend each year.

The original model for this program has been expanded in some of the competing cities. For example, in Winnipeg there are several levels of competition, starting with grades 5–6. The next level of competition is for grades 7–8 and there is another round for grades 9–10 as well as the "open" competition for grades 10–12. Some other sites also offer more than the one level.

The level of involvement of a mentor varies, though most students spend some time in a mentor's lab. Extremes range from the mentors only providing some guidance to the student with the student doing the actual experimentation elsewhere to students working with the mentor almost every day during the mentorship. Often the student works with a junior scientist, a graduate student, technologist or someone else in the mentor's lab. It is common for the mentor to provide overall direction only. This approach gives the student a wide range of mentorship involvement and often a wider experience in the area of science.

The mentor must establish his or her role in the project at the very beginning. Ideally, a mentor will have been involved in the project proposal and in the experimental design and can then establish, along with the student, the amount of time and effort that will be required to successfully complete the project. Students often have no concept of how much work is required to carry out a research project—it is not possible to carry out a successful project with meaningful results on two Saturdays. This can be a problem, as many of the students wishing to participate also lead very busy lives, being involved in sports, music, working outside of school etc. Some projects have foundered when ground rules regarding time requirements were not established up front.

3.3.3.1 Fats- and Oils-Related Case Studies

One of the authors has worked as a mentor with the SABC program on several occasions. His laboratory and expertise does not fit with the more popular view of biotechnology, i.e., gene splicing, working with microbes and utilizing enzymes, but he has been able to show students how just about any scientific exercise in the fats and oils area is related to biotechnology in the broad sense. He has always started working with students before they have developed their project proposal. In this way he has been able to ensure that the proposals developed are good science and also fitted into the general mandate of his laboratory. This was accomplished by meeting with the students early in the year, giving them a tour and outline of the work carried out in the grain research laboratory and then indicating several possible research areas for the student to develop as a project. Four of the projects are outlined below.

3.3.3.1.1 Detection of Genetically Modified Soybeans in Soybean-Based Foods

This project actually started as a part of a school-based mentor program that paired students up with mentors. The student indicated an interest in the SABC (then the Connaught program) and after discussion decided to carry out a project in which a simple enzyme-linked immunosorbant assay (ELISA)-based kit designed to detect GM soybeans would be tested for its ability to detect them in common soy foods such as tofu (soybean curd) and natto (fermented soybean paste). The student worked with the head of the Grain Research Laboratory's biotechnology division to carry out the testing and with a senior food scientist at the Agriculture and Agri-Food Canada Research Station to learn to prepare soy foods. The student discovered that processing soybeans denatured the protein so that it was not detected by the ELISA kit. In addition to placing third in the Manitoba Biotechnology Competition and winning awards in the Manitoba Schools Science Symposium, the student presented a poster at Agri-Food 2000, a major scientific conference in Winnipeg.[2] The student completed grade 12 and proceeded to technical school, where she studied mechanics.

3.3.3.1.2 Stability of Ground Flaxseed to Oxidation

Oneofus, on advice from a teacher, was approached by two students from a local high school who wished to participate in the biotechnology competition. At the time, there was increased interest in the use of flaxseed in food products and earlier research in my laboratory had shown that ground flaxseed baked in bread appeared to be stable for several months. The students decided to study the effect of fatty acid composition on the stability, comparing normal flaxseed with about 50% linolenic acid with solin (flaxseed with less than 3% linolenic acid).

The mentorship began with the students collecting samples of solin and flaxseed from a local producer and from a local grain elevator. The students then worked in the Grain Research Laboratory with a junior scientist and a graduate student. They also worked with representatives from a scientific supply company who provided the testing kits for determining peroxide value and aldehyde value. They determined that ground seed from which the oil had been extracted and replaced was more stable than ground seed and also that solin, in which the oil was replaced with flax oil, was more stable than flaxseed in which the oil had been removed and replaced. This project won awards at the Biotechnology Competition and at the Manitoba Schools Science Symposium; results were presented at a scientific conference[3] (Canadian Section Meeting, AOCS). The students involved continued as summer students at the Grain Research Laboratory. Both completed honors degrees in science and have progressed to scientific careers.

3.3.3.1.3 Inheritance of Erucic Acid in Wild Mustard (S. arvensis L.)

This project was initiated by three sisters (triplets) as a proposal to evaluate the fatty acid composition of wild mustard to see if it had been affected by canola over the years. These students wrote the original proposal and collected some samples. They determined that their sports interests would not permit them to continue with the project and it was taken over by two students from a different high school. These students rewrote the proposal to include a genetic study on erucic acid in wild mustard and carried out the laboratory studies, which included collaboration with a senior professor and technical staff at the University of Manitoba to grow out and cross samples of wild mustard and also work with the mentor and a junior scientist to test the samples for fatty acid composition.

The results of the study showed the influence of several genes in determining the level of erucic acid in this species. It also showed that wild mustard contained an unusual fatty acid series *n*-7 rather than the more usual *n*-9. This project won first place in the Manitoba Biotechnology Competition and also prizes in the Manitoba Schools Science Symposium. Results were presented at the International Rapeseed Congress in Copenhagen[4,5] and have been submitted for publication. The *n*-7 findings spurred a later project, mentored by the junior scientist, on the genetics of this fatty acid.

It was interesting that while the students in this project both worked hard and spent considerable time in the laboratory, they noted that the repetitive work was somewhat boring. One student continued his education in engineering at university and is preparing for postgraduate studies. The other student finished high school and decided to pursue a career in modeling.

3.3.3.1.4 Interesterification of Long Chain Polyunsaturated Fatty Acids with High Laurate Canola Oil

This project was initiated by a group of three male high school students. They met several times with the mentor to plan the project, which involved an attempt to form a structured lipid by interesterification high lauric acid canola oil with docosahexaenoic acid (DHA). The mentor provided the chemicals, enzyme and substrates to carry out the experiment. The students worked in both the mentor's laboratory and in their school science classroom over a period of 2 weeks. The first attempt at interesterification did not work well and the mentor proposed a second attempt with some method alterations. The students were unwilling to commit further time to the project and the project was withdrawn from the competition, although one student wrote an essay based on the project. The students have all continued on to university studying in the science and engineering areas.

Interestingly, this project later became the topic of a PhD program at Memorial University in Newfoundland.[6]

3.3.3.1.5 Future Possibilities

The areas of oilseeds and fats and oils provide excellent opportunities for student-led mentored projects. Taking current interest in the development of biodiesel fuels is an example; two potential projects might revolve around novel production of biodiesel from pressed canola cake based on USDA research.[7] The effect of biodiesel spills on the environment also requires study and this could easily be carried out as a multiple-year student project in collaboration with an environmental agency such as Ducks Unlimited.

3.3.3.2 Benefits of Mentoring

The case studies illustrate that mentoring has benefits beyond the actual mentoring process, i.e., beyond the coaching and introduction of young people to the world of science. In all of the cases above, real research was carried out that resulted, directly or indirectly, in publications or at least scientific presentations. The projects all continued beyond the student stage and became part of ongoing research that has led to significant findings. It is unlikely that these research programs would have been initiated without the contributions of the students.

Mentoring opportunities arise beyond the special instances described above. For example, one of the authors (JKD) developed a laboratory proficiency program, the income from which was used to hire a summer student to administer the program. The program administration took only about 20% of the student's time. Instead of using the student to perform basic tasks such as dishwashing etc., the undergraduate university students hired for this program were each given small research projects to carry out. Again, this provided an opportunity for mentoring students to help them to decide whether they have the desire and abilities necessary to continue with postgraduate research. Rather than being treated as a general dogsbody, the students were treated as equal participants in the research team. Students from this program have gone on to further studies and several have become either chemists or senior technicians in science organizations.

3.3.4 Desktop Videoconferencing

An interesting sidelight to the SABC program has been the use of desktop videoconferencing to bring mentoring and other educational resources to students in rural or remote areas where it would be difficult to find professional scientists with the background to challenge top students. Webcam technology is inexpensive and relatively easy to set up and use. Two technical areas have caused problems with these programs but both of these are solvable. One hurdle is the availability of broadband Internet service, which is needed to have clear pictures, voice and text. This is becoming less and less of a problem as broadband services are being offered to larger and larger areas in almost all countries, including the rural and remote areas of industrialized countries like the United States and Canada with their enormous land mass. The firewalls of school computer networks can be another issue. Firewalls are constructed to keep unwanted intruders out of the network. Sometimes these network security tools perceive the incoming videoconference information as unwanted and will block the signals. This has been handled by working with school technology coordinators to open a "port" for the videoconference to come through.

In addition to providing mentors a tool to communicate with interested students, videoconferencing can provide support for teachers who don't have enough background in science. Schools in rural areas tend to be smaller than urban schools, reducing the specific expertise in any given course because educators must teach a larger number of subjects, including ones for which they are not specifically trained. Further, schools in remote areas may have only one or two teachers responsible for all grades and subjects, K–8. In these situations, there is no guarantee that teachers will have the time or expertise to teach specialized subjects such as science. Videoconferencing has been used by MindSet to deliver this specialized information.

3.3.5 Agricultural Biotechnology Enrichment Program

Agricultural Biotechnology Enrichment Program is a cooperative effort between Monsanto Canada and the Pembina Trails School Division, under the direction of Bob Adamson, an experienced science teacher at Fort Richmond Collegiate in the Fort Garry suburb of Winnipeg. Adamson approached Monsanto about creating a portable laboratory that would introduce agricultural and medical research. The laboratory, used by many schools, is shipped in large rubber footlockers, containing:

- Micropipettes
- Power supplies
- Gel electrophoresis chambers
- Gel cameras for documenting experiments
- Polymerase chain reaction (PCR) technology

Teachers from schools that want to make use of one of the three footlockers must undertake the training necessary to make the best use of the technology. This training becomes an extra facet to the teachers' professional development. Since its inception in 2001, this program has had more than 1,800 students participate. One of the strengths of the program is that it was developed to match the learning outcomes of Manitoba Education, Citizenship and Youth, the province's Ministry of Education.

3.3.6 Community-Based Programs

We were also involved in several interesting cooperative projects in which the outreach programs of science and technology organizations use their proximity to schools as a leverage point to engage local communities in science. These programs are based on the example of a biotechnology lab at Ho Yu College in Hong Kong, a K–12 high school. Dr. William Mak, director, Genome Research Centre, Hong Kong University, was a pioneer in biotechnology education in Canada. After 35 years in Canada, he had an opportunity to return to Hong Kong in his present capacity. He also became president of the Hong Kong Biotechnology Education Resource Centre. He worked with Suet Ying Lee, principal of the school, to build a state-of-the-art biotechnology lab on her site to first train her teachers and students in biotechnology, followed by training teachers from other schools. Once the teachers were trained, they could bring their own students there.

This model is being explored in Winnipeg for three labs possibly being built in schools close to research facilities. The cost of the labs will be funded by the research institutions and their partners. A part-time lab manager will coordinate the programs. Researchers from these facilities will supplement the hands-on training being provided educators and their students so they can go into "real" science. These researchers can do presentations about science, in general, or demonstrations related to their own work.

3.3.6.1 Business of Science Student Stream: Involving Students in Professional Events

Another way of providing students with experience about science and scientists is to invite them to professional events. Each year since 2003, the Life Sciences Branch of Manitoba Energy, Science and Technology has invited 50 high school students and an equal number of college and university students to the Business of Science Conference, a prestigious international conference held in Winnipeg to help scientists better understand how to deal with the business side of the equation. The secondary and post secondary students each have their own sessions, as do the professionals participating in the event. During keynotes, meals and session breaks, the students and their teachers have an opportunity to mix with scientists. In addition to providing an opportunity for students to realize that scientists are pretty regular people, these networking possibilities can lead to partnerships for schools. Mentorships, internships or job-shadowing activities for students have also grown out of these conferences. In fact, the same model is used for information and communications technology, new media and advanced manufacturing conferences.

3.3.7 Gene Researcher for a Week

Since 2002, the Canadian Genetic Diseases Network, in collaboration with the Canadian Institutes of Health Research and Merck Frosst, has invited students to apply to work in a genetics laboratory during their spring break. The lab might be local or in another province, but students must compete for one of the prized 30 spots offered nationally. Expenses are paid for travel to and from the lab as well as for the time spent in the laboratory.

3.3.8 Harnessing the Grey Matter

Harnessing the Grey Matter was a program to match scientists interested in working with schools with schools interested in having professional support for their science programs. The program was developed by the Winnipeg chapter of the Sigma Xi Society in partnership with MindSet, with a grant from the Promoscience program of Canada's Natural Science and Engineering Research Council over a 3-year period. After that time, the program was taken over by the ministry responsible for education, Manitoba Education, Citizenship and Youth, and renamed "Scientists in the Schools."

The "grey matter" in this context was a play on words that related to the "greying" of the population of scientists in Canada as well as their brains—their "grey matter." The original direction of the program was to identify retired scientists or those close to retirement as science resource people for schools. As people, scientists tend to be both interested and interesting, and also very passionate and involved with what they have been doing. This passion keeps many of them involved in their research and other aspects of science even after retirement. Harnessing the Grey Matter wanted to engage this devotion and channel it into educational activities. Among the possible tasks for these scientists on the program's web site were the following:

- Visiting-scientist programs
- Professional development of science teachers
- Student mentors
- Science advisors to schools
- Media-contact persons
- Writers and journalists
- Science fair judges
- Conference speakers

Upon hearing of the program, scientists, engineers and technologists not close to retirement also registered on the web site so that schools could find them in the database.

3.3.9 SMARTS: Of the Students, For the Students, By the Students

The Student Mentorship Association Regarding Technology & Science (SMARTS) is something completely different from anything in this chapter. The program to support and promote Science, Technology, Engineering and Mathematics (STEM) to students in grades 7–12 was founded by students and is still run by students as a program within Youth Science Fairs Canada. The founding students have been very successful in science activities, mainly science fairs, in their own provinces. Their participation in the formation of SMARTS was altruistic—to give something back. The purpose of the organization is to inform other students about various STEM opportunities and to inspire them to get involved. At the same time, SMARTS strives to provide resources and support for these other students. Their main vehicles for this are the school correspondent network, an events, programs and competitions directory and a compilation of helpful suggestions for students competing in science competitions.

3.4 WHAT ARE THE BARRIERS?

There are issues, however, serious ones at that. Schools have been increasingly forced to follow curriculum topics dictated by their local jurisdiction at a state, provincial or school-district level. At the same time, if these documents do not contain topics related to biotechnology as an important area of change in agriculture, the topics might not get taught even if teachers are interested in teaching them.

The profile of biotechnology remains low in the mainstream of most people's lives—including teachers, apparently. For example, a recent survey in Canada showed that most Canadians had no reaction to the term "biotechnology" and 90% of them did not know that Canada was a world leader in the area and were unaware of the economic benefits of biotechnology. Among teachers, the profile of biotechnology does not seem much better. Canada's first national bioscience and biotechnology conference, held in Winnipeg in July 2006, attracted only 34 participants, in spite of national marketing. The objective of the conference, which showcased tours of world-class facilities and presentations by international and national scientists and educators, was to provide educators with teaching ideas to increase the amount of biotechnology in their courses. The previous month, a national contest to identify Canada's outstanding biotechnology educators produced only a small handful of nominations.

3.4.1 Involving Educators

Educators—teachers, principals, consultants, directors—are an integral part of the process of developing a science culture. As noted elsewhere, they are the gatekeepers to getting students involved. Teachers can be very busy people, with preparing lessons, marking student work, duties assigned at school and the "volunteer" activities they take on as coaches, tutors and managers of various student activities such as drama, music and other cultural events. Such a busy schedule leaves little time for staying abreast of changes in their particular areas of expertise, especially if they are involved in science areas with high rates of change.

In addition to organizing many student activities, MindSet provides educators with support for their science programs. The Life Sciences Educators Group has been developed to provide easier networking to the business sector and postsecondary institutions. Professional development sessions about cutting-edge science are also organized on a regular basis, most often using speakers who are leaders in their field.

3.4.2 The Voice of Scientists

Science and technology are constantly in the media. Some of that information is good; some of it is bad. However, the voice of the scientist is not necessarily heard as part of the dissemination of information. That voice would echo the feelings of many scientists, who are driven by their passion for their research, their desire to contribute to the common good and their desire to know "why." More scientists need to share this aspect of their efforts. The message, of course, needs to go to the community, but it especially needs to be heard by young people as they consider how they might spend the rest of their lives.

This evangelizing would come easily to many scientists. The qualities that make good scientists don't necessarily make good public relations people. Nonetheless, there are other alternatives to pushing someone else off the soapbox. For example, the Manitoba Science and Technology Achievement Lunch is set up each year so that students and teachers have the opportunity to meet practicing scientists, engineers and technical people. The event has attracted as many as 375 people whose contributions or achievements in science and technology have been recognized at a provincial, national or international level. While the event is only a lunch, followed by an interesting speaker involved in some area of science, it is certainly an opportunity for scientists to share their knowledge with students and their teachers.

REFERENCES

1. Daun, J.K. Students don't get enough fat—in the classroom, *Canadian Chemical News 2002*, 5, 2002.
2. Limieux-Robinson, I.; Giroux, R.; Daun, J.K.; Brown, B.; Scowcroft, B. Determination of the presence of GMO in soybeans in soyfoods, Agrifood 2000 Conference Winnipeg, 58. 2000.
3. Botha, I.; Hilderman, E.; Daun, J.K.; Przybylski, R. Protective effect of meal on lipids of flax and solin *(Linum usitatissimum), Program and Abstracts of the 16th Annual Meeting of the Canadian Section of the AOCS,* Winnipeg, 2001.
4. Daun, J.K.; Barthet, V.J.; Scarth, R. Erucic acid levels in *Sinapis arvensis* L. from different parts of the world, *Proceedings of the 11th International Rapeseed Congress.* Copenhagen, Denmark: The Royal Veterinary and Agricultural University. 290–292, 2003.
5. Scarth, R.; Daun, J.K.; Barthet, V.J.; Nugent-Rigby, J. Inheritance of erucic acid content in European and North American populations of wild mustard *Sinapis arvensis, Proceedings of the 11th International Rapeseed Congress.* Copenhagen, Denmark: The Royal Veterinary and Agricultural University. 293–295, 2003.
6. Haman, F.; Daun, J.K. Lipase-assisted acidolysis of high-laurate canola oil with eicosapenaenoic acid, *J. Amer. Oil Chem. Soc. 82*, 875–879, 2005.
7. Haas, M.J. New method simplifies biodiesel production, *Agricultural Research* April 13, 2005.

4 Teaching Lipid Chemistry through Creative Problem-Solving

Karen M. Schaich

CONTENTS

4.1 INTRODUCTION

Teaching any science is a challenge because fundamental concepts are not intuitive for most students and explanations are not easy to connect to real experiences. Hence, "facts" are memorized for exams and, with luck, most students will remember a little of the information a few weeks after the exam. Hopefully, some may remember enough to use in later classes.

Students can learn to think and reason intuitively if they are introduced to and given practice in the processes involved in transforming data or information to actionable knowledge.[1] Research with learning has shown that long-term memories are generated when information is connected to something already known—i.e., new neuronal paths are connected to existing neuronal and memory networks in the brain.[2] Learning research has also shown that, in contrast to long-time dogma, ability to think theoretically was entirely innate, the ability to cycle between concrete (which everyone comprehends) and theoretical concepts (which relatively few really understand) can also be developed:[1,3,4]

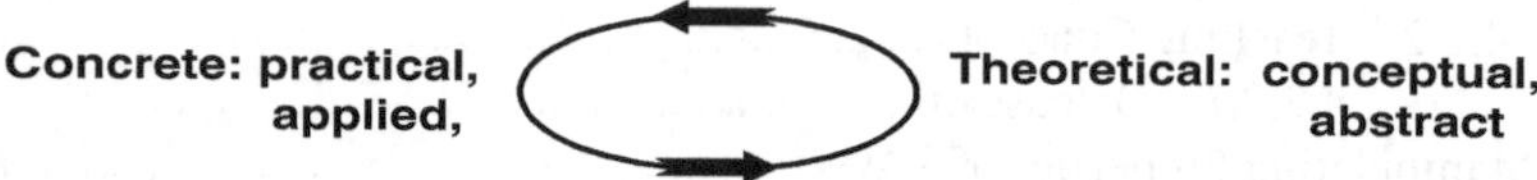

Considering this research and analyzing courses in which I *learned* versus those in which I just *memorized* has led me to teach my graduate classes, especially Lipid Chemistry, in a problem-solving context designed to connect fundamental concepts to real-life recognizable situations. In the education lexicon, this is a modified case-study approach. The first class on each topic is a lecture covering new and critical material, and then students are given several days to study and think about the material (or an assigned problem). The second class is a problem-solving session presented as individual research/thinking exercises prepared at home with open discussions and debate in class, spontaneous (no prior exposure to problem) exercises in small cooperative learning groups or directed learning as a formal case study with the entire class working together. Sometimes everyone is given the same problem; other times, multiple problems covering different aspects of the topic are distributed within the class, prepared solutions are presented and the rest of the class debates the issues extemporaneously. Discussions are augmented by questions designed to elicit prior knowledge (students know more than they think they know, but don't know how to connect data to action), stimulate exploration for alternatives, induce evaluation and develop logic in thinking processes, understanding and problem-solving strategies. Submitted written answers are scored by a rubric designed as a teaching tool to recap key points that should have been recognized or issues addressed, and also award major extra credit for new ideas and independent thinking. Students can use the rubrics to build and improve their problem-solving strategies and critical thinking through the semester. Significant points toward the course grade (e.g., 15%, enough to make a grade difference) are awarded for class participation to encourage active student involvement in discussion and debate and to provide a reward for students "taking a chance" with tentative or new ideas, or even interesting questions that show thought, creativity and connection.

The problem-solving sessions provide an environment where students can learn by exploring, trying out ideas and making mistakes without fear or threat. The only "stupid" question is one that is not asked; the only "bad" idea is one that is not voiced and considered thoroughly. Regardless of the problem format, each student contributes ideas, and the class evaluates whether the idea will work or why not—and if not, how to fix it. Although each student in most cases prepares an independent proposed solution to submit, the composite solution developed in class ultimately results from collective efforts and reflects multiple perspectives. Critical concepts are stressed, usually more questions are asked than set answers given, creative thinking is encouraged and connections between fundamental concepts and real-life behaviors are developed and reinforced. Problems early in the semester are relatively straightforward while students learn how to think through solutions. Subsequent problems build in complexity as students gain more information in lectures, learn how to apply and integrate what they already know with in-depth technical material from lectures and build a more extensive base of critical knowledge and understanding. Typically, students feel uncomfortable during the 4 to 6 weeks, but reach a "eureka" point about two-thirds of the way through the semester, where everything suddenly comes together. Then they are able to both recognize the science controlling a situation and apply principles creatively and with depth to propose workable solutions.

In addition to teaching thinking and learning strategies, problems can show students quite graphically how *details count*. *Details* provide power in an analysis. *Details* make processes work. Students are so accustomed to learning generalities that they fail to recognize that the power and control of a situation is in being able to use details to advantage to provide context, a frame of reference, and depth of understanding. For example, students readily associate saturated or long chain fatty acids with high melting points, but what is "high"? Is it the temperature of a warm room, the kitchen counter in the sun, human bodies, boiling water or something even higher? Which fatty acids melt under each of these conditions? What does that difference mean in terms of observable behaviors? Fatty acids are generally not volatile, but volatilization can be forced under vacuum or after conversion to esters. This is an easy concept to understand, but details are needed to apply it to determine temperature-pressure conditions required for physical refining or the temperature gradient that will separate a specific fatty acid mixture by gas chromatography. Similarly, aldehydes and ketones are well-known products of lipid oxidation, but what does a shift of product mix, e.g., from hexanal to heptenal or octenal, tell about the mechanisms and conditions of oxidation, and how does this affect decisions about control strategies? Working through problems as a group provides many opportunities to wake students out of their mental sleep and turn them on to the power of details, not as meaningless lists of numbers stored away in references, but as critical tools essential to the success of any proposed application or problem solution.

I use problem-solving intensively to deal with complex material in advanced graduate courses such as Lipid Chemistry, but problem-solving can be adapted to any level, even elementary through secondary school (K–12), by changing the nature of the questions and complexity of the problems. Undergraduates and new graduate students, for example, can be challenged to recognize the basic science underlying phenomena in everyday life or observations in research papers. K–12 students love to see how science can make sense of what they see every day. In all cases, the goal is to move students from merely recalling memorized facts to thinking beyond the information or words of the problem and using what they have learned to understand their world—analyzing the scientific driving forces and applying their knowledge to gain control of the situation.

To support development of problem-solving and thinking skills, students are given three tools: Schaich's Law, the Active Learning Process and a guide with Problem-Solving Strategies. Schaich's Law (Figure 4.1) states that: "Foods are not black boxes—they behave according to fundamental laws of chemistry and physics. If you want to understand and control food properties and characteristics, you must first identify the underlying chemistry responsible for each property of interest. Then you can use and manipulate that chemistry to your advantage." This law is a constant reminder to students that foods are chemical systems; it pushes them to think about foods as (collections of) molecules rather than macroscopic systems. Whatever is seen happening in foods can be explained in terms of underlying chemistry.

From an early age, students are drilled in memorizing and regurgitating facts, but they are often stymied when asked to extend, apply or connect those facts to real situations. The Active Learning Process (Figure 4.2) provides a road map to move students past rote memorization of facts to higher-order thinking, make them aware of the complex processes involved in learning and encourage them to make learning a mindful, conscious process.

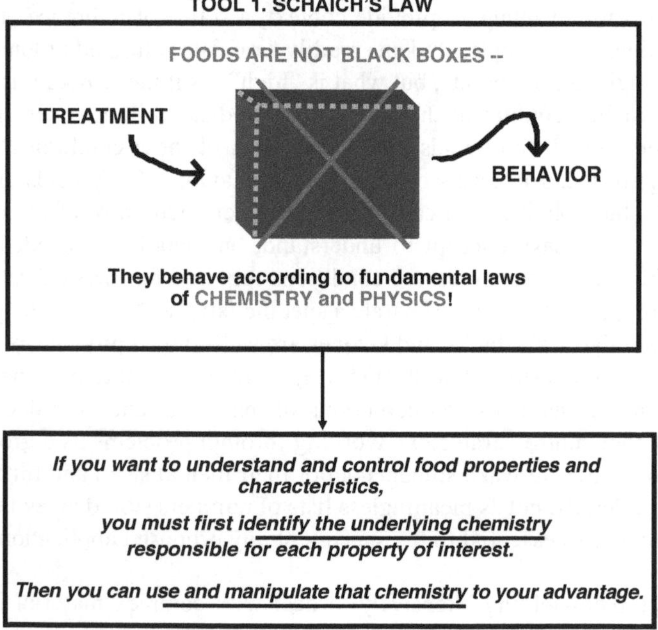
TOOL 1. SCHAICH'S LAW
FOODS ARE NOT BLACK BOXES --
TREATMENT
BEHAVIOR
They behave according to fundamental laws of CHEMISTRY and PHYSICS!
If you want to understand and control food properties and characteristics,
you must first identify the underlying chemistry responsible for each property of interest.
Then you can use and manipulate that chemistry to your advantage.

FIGURE 4.1 Tool 1. Schaich's Law reminds students to think of foods as molecules rather than macroscopic systems.

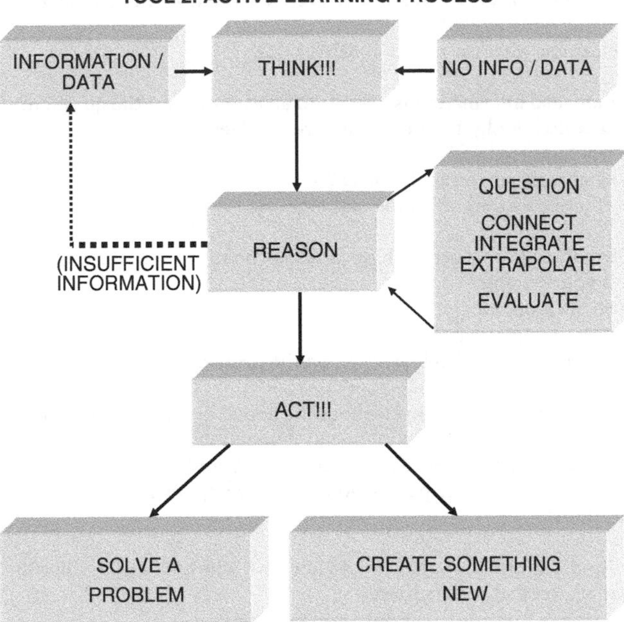

FIGURE 4.2 Tool 2. The Active Learning Process makes students aware of the complexities involved in learning and encourages them to make learning a mindful, conscious process.

Figure 4.3 provides a starting point for thinking, not a checklist for answers. These questions guide students to probe beyond obvious and general answers, to think in greater depth, to apply details specifically and selectively, and to explore how what they do not know may be important to the solution. In most cases, what is not known at one stage is likely to be covered later in the course, so discussion of "unknowns" builds anticipation and provides a framework for recognizing applications of fundamental information in lectures to come. Perhaps most important, the process outlined in the strategies, with minor rewording, is broadly applicable to any problem and, if adopted, can change the way students think in their jobs and daily lives. In educational parlance, such learning becomes *transformative*.[5–7]

This chapter presents a representative sample of problems used to teach Lipid Chemistry. Most are aimed at the advanced graduate level, but one emulsions problem that can be adapted to any level, including K–12, is also included. These "problems" have been collected from a variety of real-life situations—industry problems, government agency challenges, research papers, communications from colleagues, personal observations—wherever something with interesting possibilities for teaching pops up. Due to space limitations, only a few problems will be discussed in detail to show how they are used to reinforce basic concepts of Lipid Chemistry and to develop higher order thinking and problem-solving skills; others will be listed along with the teaching concepts covered in each. Problems have been designed to coordinate with, reinforce and support each Lipid Chemistry topic covered in lectures (Table 4.1).

TOOL 3. PROBLEM-SOLVING STRATEGIES

1. Decide/determine what the question REALLY is.

2. What fundamental information is needed to answer or solve the question?
   What chemical or physical properties are involved?
   What reactions are involved?
   What system properties are involved?
   What other issues may affect the outcome?

3. What information do you already have that may be relevant to the problem?
   What is your system?
   What are its properties?
   What is the environment?
   What resources are available to work on the problem?

4. What does this information tell you about the actual or potential behavior in the system under question?
   How do your system properties control its behavior?
   What can you expect to happen if you change components or conditions?

5. What additional information that you do not have will you need to solve the problem?
   Where can you get that information --
   Books and publications
   Experts in the field
   Suppliers
   Research

6. How can you fit everything together creatively to solve the problem?
   Don't overlook the obvious and conventional approaches, but also always try to *THINK OUTSIDE OF THE BOX!*

FIGURE 4.3 Problem-Solving Strategies: A Series of Questions That Guide Students in Thinking through Problems and Developing Individual Approaches to Analyzing Situations and Developing Solutions

TABLE 4.1
Lipid Chemistry Concepts Taught at Various Levels Using the Problem-Solving Approach

PhD Level
General lipid classifications, structures, characterization, properties and functions
Chemical and physical properties of fatty acids
Chemical and physical properties of triacylglycerols
Chemical and physical properties of phospholipids
Processing of fats and oils for food use

(Continued)

TABLE 4.1
(Continued)

- Lipid structures in foods and biological materials
- Characterization and analysis of lipids: Chemical analyses
- Characterization and analysis of lipids: Instrumental analyses
- Emulsions and emulsifiers
- Degradation reactions of lipids—Heat
    - Radiation
    - Enzymes
    - Photolysis
    - Autoxidation—Basic reactions
        - Pro-oxidants
        - Antioxidants
- Health issues of lipids; fat substitutes
- Hot topics in lipid chemistry

Undergraduate and MSc Level

- Classes and structures of lipids—derived, neutral, complex
- Lipid functions in foods
- Physical properties of lipids and their relationship to lipid functionality
    - Solubility
    - Viscosity, plasticity and consistency
    - Melting point
    - Solidification/crystallization
    - Flavor properties—carrier, potentiator, component, precursor
- Reactions of lipids
    - Modification (chemistry and applications)
        - Hydrogenation
        - Esterification (intra, inter)
    - Degradation reactions (chemistry and applications)
        - Saponification
        - Hydrolysis
        - Heat
        - Enzymes—lipase, lipoxygenase
        - Oxidation—basic reactions and conditions, catalysts, antioxidants
    - Emulsions and emulsifiers

Concepts Introduced in Materials Developed for K–12

- General structure characteristics, intermolecular associations deriving from structure relationship of structure and associations to properties of fats, alone and in foods
    - Solubility
    - Viscosity
    - Melting/solidification
    - Crystallization and solid fats
- Emulsions—General properties and emulsifiers
    - Applications in mayonnaise and ice cream

4.2 EXAMPLES OF PROBLEMS

4.2.1 Making Margarine from Oil

As product development manager for an entrepreneurial food company specializing in corn utilization, you have been directed to produce a line of margarines that includes traditional stick margarines as well as tub margarines and new pourable margarines based on corn oil. You have one week to give the board a development plan that includes your proposed production approaches as well as the scientific justification for them.

Points to consider in developing your plan:

(a) What physical or chemical properties of lipids (oils) are needed for each type of margarine?
(b) What are the endogenous properties of the corn oil used as a base? How must these be modified to produce each type of margarine?
(c) What process(es) can be used to modify oil properties?
(d) Outline several approaches that could be used to produce stick, tub and pourable margarines from corn oil. Be specific about the steps to be used and what each step will accomplish. Include scientific justification for your proposed methods.
(e) Discuss advantages and shortcomings (if any) for each approach.
(f) What information are you missing to solve this problem completely? Where could you find this information?

4.2.2 Teaching Concepts

Introduction to margarine processing

Identification of target properties for each margarine (mp, consistency/texture/other)

- Likely components, etc. associated with these properties

Factors controlling melting points and melting behavior:

    Fatty acid chain length and degree of unsaturation
    Geometric and positional isomers of fatty acids
    Arrangements of fatty acids on glycerol backbone
    Alignment and packing arrangements of fatty acid chains

- Distribution of different types of fatty acids in mixtures

Factors controlling solid properties

    Melting points (degree of saturation/unsaturation)
    Crystal structures—polymorphism
        Packing arrangement of fatty acid chains in pure fatty acids, in triacylglycerols
        Specific mix of fatty acids

- Characteristics of semisolids—viscosity, consistency, mouthfeel, plasticity

- Concept of solid properties as a broad range gradient rather than step function

Fatty acid composition of corn oil and melting points of component fatty acids

Approaches to obtaining target properties for each margarine

Change mp—hydrogenation

Change crystal structure—intra/inter-esterification

Change both and avoid unwanted side reactions by blending with other oils

Pros/cons, advantages/disadvantages, controversial issues of each approach

Identify reaction control as unknown

Integrating information into solutions and strategies

4.2.2.1 Discussion

This problem is typically presented with the initial fatty acids lecture and is used as an introduction to problem-solving processes. Hence, it is relatively uncomplicated and includes more questions as learning tools to guide students in thinking beyond the obvious. In trying to teach a logical mental approach to applying information, the starting point is defining the problem and outlining critical factors involved in the starting materials, the process, the problem and the solution. Even at the undergraduate level, all students can identify melting point as the controlling factor and that high, medium and low melting points are needed for the three margarines. The key to solving the problem is being able to apply a variety of information about melting points to the specific materials. Here is where details count.

At this point in the semester, all students will recognize fundamental relationships and reactions, but few will go further to look up specific data, e.g., melting point ranges for each type of margarine, the fatty acid composition of corn oil, melting points of component fatty acids—both free and in triacylglycerols—and melting ranges attainable with various degrees of hydrogenation (information that is in fact in the course materials), information that will be critical to produce the final margarines. Similarly, although students understand general concepts of melting points and crystal structures, they have difficulty in integrating the two to generate solid properties. In-depth discussions with considerable probing provide a framework within which students begin to understand interrelationships among melting behavior, crystal forms and texture properties in real products, and begin to see that details of oil composition and fatty acid behaviors are critical for defining their target properties.

Considering how melting points can be altered theoretically then leads to ideas about how melting points can be modified practically. Hydrogenation is introduced in the fatty acid lecture as a reaction of double bonds, and all students have had some introduction to lipids and lipid reactions in a prerequisite food chemistry course. The idea in the discussions is to extend knowledge about a basic reaction to create a process, or to recall prior general knowledge about a process and apply it to a specific situation. Here, *unknowns* become important. Students may recognize that oil properties can be modified by hydrogenation and inter- or intra-esterification, but ask them *how*? or *how much*? or *how do you make A versus B*? and they are stymied. Information that is unknown or not understood often is the factor preventing learning or development of successful solutions to problems, but unknowns are seldom recognized or evaluated. Hence, analysis of knowledge structures and identification of unknown information is stressed in all problems throughout the semester. Details of hydrogenation and inter-esterification reactions and methods for controlling these reactions to tailor product properties will not be introduced until later in the semester. However, this beginning problem shows quite clearly how critical control of the reactions is for attaining desired specific properties, and thus provides a reason for students to look forward to later sections. It also establishes a broad and firm base to which new information can be connected and where technical details will make sense and be important rather than overwhelming. This simple margarine problem is a building block linking fatty acids to triacylglycerols and then to processing and even analyses, with each new lecture adding to this base and increasing depth and breadth of understanding.

Anticipation of future lectures can be further enhanced by asking questions about analyzing the progress of hydrogenation reactions, determining triacylglycerol (TAG) structure generated by inter-esterification, health effects of fats/oils and their modified versions, regulatory issues, oxidative stability and antioxidants, etc. Thinking about these questions, even without answers, begins to create the logical connections from which knowledge networks will be built throughout the semester.

It should be noted that successful use of problem-solving requires considerable involvement of the professor in guiding—not directing—class discussions in both process (supporting students with reminders, hints, questions, encouragement) and content (amplification or more detailed explanation of basic concepts, introducing related new issues). This is scaffolding, or building a learning support framework.[8,9] The semester starts out with the teacher's having most control and guiding students into new thinking processes. As the semester progresses, students learn more about lipid chemistry and how to reason with it and use it; they gain both confidence in their abilities and competence in their assessments, and they play a larger role in directing class discussions. As this happens, the professor role shifts increasingly to contributing experience, expertise and perspectives and add depth and breadth to the exercise.

4.3 MANIPULATING PROPERTIES OF TAGS

1. You are an Oompah-Loompah working at Willie Wonka's Chocolate Factory and have decided that the Augustus Gusses of the world can be just as satisfied with chocolate candies made of cocoa butter equivalents (CBEs), which are much cheaper, as with expensive hand-made real milk chocolate. Willie Wonka says, GO FOR IT.

   (a) What lipid and final product properties or factors must be considered in developing cocoa butter equivalents?

(b) From the information about fats and oils you have in your course materials, select two candidate fat sources for the CBE, one of which must be a non-tropical fat.
   - Explain why you chose these fats.
   - What modifications will be necessary for each?
   - Suggest detailed approaches for development of the CBE using each fat.

(c) What information that you need to solve this problem are you missing?

2. You have just become the general manager for the Sweet-n-Sour Dairy Cooperative that can't supply skim milk to consumers fast enough because all the space in the plant is being filled with the butterfat removed from the milk. Because you are an optimist, you believe that your opportunities for growth and remuneration are as full as your plant.
   (a) What are the general chemical and physical properties of butterfat?
      - Which have positive attributes (give the context)?
      - Which have negative attributes (give the context)?
   (b) Which attributes do you think offer some opportunities for developing alternative products from butterfat? Why? (These applications do not have to be food applications.)
   (c) Suggest approaches for utilizing the butterfat.
   (d) What information that you need to solve this problem are you missing?

3. Floods in the Midwest destroyed the corn and soybean crops that produce source oils for several products your company manufactures. At the same time, excessive and unusually dry weather has limited supplies and drastically driven up prices of coconut, palm and cocoa fats/oils, which your company uses in other products. As product development manager, you are directed to find substitute oils or fats. Your board of directors also decides to avoid similar disasters in the future by expanding operations vertically, i.e., adding oil production and refining to company activities.

Using the data in your course packet about the properties and fatty acid composition of natural oils and fats, or any other information you can find, propose alternative oil sources:

(a) To *buy* to replace presently used corn, soy, and palm/coconut oils and cocoa butter. Include your reasoning and justifications.

(b) To begin to grow for the following applications. Include your reasoning and justifications.
   - Substitute(s) for corn and soy in high volume products
   - Substitute(s) for corn and soy in specialty products with limited markets
   - Cocoa butter replacement
   - Replacement source for coconut and palm

(c) What information needed to solve this problem are you missing?

4. You are a drone for Busy Bee Bakers, who want to corner the cookie market with a sandwich cookie containing "healthy" fats. You recognize that consumer acceptance will critically depend on maintaining the typical crunchy texture of the cookie and the smooth, melting consistency of the cream fillings. Also, there must be no off-flavors. Furthermore, the cookies must survive greedy, grasping toddlers' hands, squashing in book bag and sharing lunch boxes with fruit and juice.
   (a) What chemical and physical properties or characteristics will be necessary in the specialty fat(s) you will use for the cookie? The filling? Why are these properties important? Which can be modified without negatively affecting consumer acceptance of the cookies?
   (b) What health issues will you address in formulating the cookie and filling?
   (c) What fats and oils will you use for these two applications? Why did you select them? What modifications will be necessary?
   (d) Suggest approaches for the cookie development.
   (e) What information needed to solve this problem are you missing?

4.3.1 Teaching Concepts

- Fatty acid composition of natural fats/oils and products derived from them (e.g., chocolate, tropical fats, conventional vegetable oils, specialty oils).
- Specific chemical and physical characteristics of individual fats or oils.
- Common characteristics and substitutability of different oils.
- Melting points, solid properties, crystal forms, plasticity of TAGs.
- Plastic vs. nonplastic melting behaviors.
- Practical implications and creative application of TAG properties.
- Functional properties of fat or oils in traditional and nonconventional food products—looking beyond traditional ingredients.
- Fractionation of fatty acids/TAGs by chain length and unsaturation.
- Modification and analysis issues.
- Market forces and product costs.
- Integration of health issues with functionality.
- *Chemistry* solves "problems" in industry.

4.3.1.1 Discussion

TAGs are so varied and have so many different sources, properties and applications that it is difficult to cover enough in a single problem. A more interesting and revealing approach is to have different students or different groups working on a series of problems that cover a variety of applications of TAG chemistry, physical properties and functions. Presentation and discussion of all the problems in class gives each student exposure to every problem, time to think about and fully develop one problem and an opportunity to apply what they know to solve others "on the run" in class. In comparing and contrasting other problems with their own, students begin to see commonality in some TAG behaviors and can identify key properties that are specific to individual oils. All together, these activities provide students opportunities to use multiple levels and types of thinking as well as different types of mental processing in finding solutions to the problems.

Making connections between oil properties and specific product behaviors is a major emphasis. As with the margarine, students see the connections superficially but not yet in enough detail to develop control over formulations or processing to create specific product qualities. The chocolate problem comes from real life. All students have eaten cheap holiday chocolate bells and eggs targeted at children who go for the taste and discriminate texture poorly, so they have personal experience for reference. Similarly, producing high-quality cocoa butter substitutes or equivalents is an important economic problem in industry. What makes cocoa butter unique? Is it really possible to re-create another fat or oil in the image of cocoa butter? What controls or modifications are required? What are the advantages or product quality losses in doing this? Discussions range from philosophical or ethical (if, as the Aztecs thought, cocoa is the food of the gods, why should we try to change it?) to theoretical (what makes the TAG structure in cocoa butter so unique?) to practical (how can the cocoa butter TAG structure be duplicated by controlled inter-esterification? What other fats have similar properties or can be modified to create them?). The concept that *details count* is stressed again as students are sent to look up detailed information about CBEs in industry and about properties and fatty acid compositions of alternative fats. It should be noted that the course packet contains extensive background information about natural fats and oils, TAG properties, specific TAG applications and current market issues. Students are encouraged to use these materials as starting points, then go to the web and research literature to fill in gaps and deepen their understanding. Importantly, this one problem probably generates more interest in inter-esterification mechanisms and control than any other application, so it serves as a very compelling positive inducement to learn the complicated chemistry of inter-esterification in subsequent lectures.

The butterfat problem introduces prospective use of information. If you have a lipid with a given set of properties, what can you do with it? Students need to use this kind of critical creative thinking in industry as much as retrospective solving of an established problem. Practice in this kind of application is very important because it requires an organization of thinking processes different from that which students normally use. At the same time, it forces students to think about how fatty acid properties change with chain length and unsaturation and how these properties can be used differentially. It also introduces issues of isolation and analysis as unknowns (again, to be covered in later sections) and of real-life industry problems that need creative solutions.

The idea for the flood problem came from one year in which there were disastrous weather conditions worldwide and markets for major crops and natural materials were extensively disrupted. Weather uncertainties are a fact of life with natural products, and the fats and oils industry and food manufacturers must be able to adapt. This scenario provides a perfect backdrop for comparing properties of different source fats on many levels—compositional and functional properties, growing conditions, market forces and prices, possible political and regulatory issues, etc.—and evaluating trade-offs in use of alternative fats. Considering issues beyond the chemistry of lipids is important to industry, where most students will make their careers, and it also stimulates different types of thinking and right-brain analyses, in contradistinction to the typical quantitative analytical analyses they normally use.

The cookie problem introduces TAG shortening and confectionary functions that are important in the food industry. It was derived from personal communications from a scientist involved in the original reformulations of Oreos when health effects of tropical fats were questioned, and is designed to force students to connect functional properties to fatty acid composition and physical properties, to consider alternatives and trade-offs in formulations and to introduce health issues in the use of fats and oils.

As with the margarine problem, these applications of TAG chemistry provide numerous opportunities to introduce additional issues, e.g., modifications, analyses, stability, health effects, economic or market considerations, regulatory issues, etc., that will not be covered until later in the semester—another taste to whet the appetite and provide connections to future lectures.

4.4 DESIGNING A USE FOR PHOSPHOLIPIDS

You are a research scientist in a multinational conglomerate that has divisions in chemistry, foods, pharmaceuticals and environmental protection. While working in the laboratory, you discovered a new class of molecules that you named phospholipids, and you characterized the physical and chemical properties of major compounds. However, you won't get your bonus unless these new compounds make money for your company.

Carefully consider the physical and chemical properties of phospholipids. (The information contained in your class notes is what you determined in your laboratory.)

(a) Propose at least one application for which your company could develop and market a product based on phospholipids.
(b) Explain your rationale (scientific reasoning) for the application, e.g.:
   - Why do phospholipids appear promising or feasible for your application?
   - What fundamental chemical properties or reactions are being utilized?
   - How do the specific properties relate to the application?
   - How does extension/application of those properties create the application?
(c) Identify information you are missing and need to find or get help with to fully develop your proposed idea.

The product or application can be in any field already covered by your company, or may branch into a new area that would offer new opportunities for the company.

4.4.1 Teaching Concepts

- Prospective application of information to CREATE in contrast to retrospective use of information to solve existing problems.
- Physical and chemical properties of phospholipids
  - Amphiphilicity and surface activity
  - Water-binding/hydration
  - Metal binding
  - Formation of bilayer structures—mesophases, liposomes, membranes
  - Liquid crystals
  - Associations with other molecules
  - Anti- and pro-oxidation effects
- Connections of chemistry to physiology, medical, environment, real life
- Identification of missing information needed to make an idea "work"

4.4.1.1 Discussion

Like the butterfat utilization described above, this problem is designed to stimulate a new kind of thinking—prospective application of information to CREATE rather than retrospective use of information to solve problems after the fact. Using guidelines established in creative problem solving or brainstorming, no idea is rejected because it "won't work" or "isn't good enough," etc. Instead, the scientific basis of each idea is examined. Phospholipid chemistry is discussed in the context of each proposed application; feasibility is evaluated in terms of advantages, disadvantages and what will be required to make an idea really work. In the process, students once again learn how details really count.

4.4.1.1.1 Examples of Applications Proposed by Students

- Dispersants for dry soup mixes
- Cold-water dispersant for chocolate drink mixes
- Improve solubility and rehydratability of freeze-dried fruits
- Oil herder for oil spills
- Stabilize, bind water and control phase-inversion of low-fat emulsions
- Combine with lipase to maintain animal fat emulsions (e.g., sausages) and increase yield of breakdown products for flavors
- Acceleration of lipid oxidation in milk fat globules for flavor generation
- Optimization of liposome formulations as drug carriers
- Cholesterol deposit cleaner to dissolve and solubilize deposits in arterial walls
- Binders for control of circulating arachidonic acid (prostaglandin control)
- Natural antioxidant in vegetable oils
- Use with C16/C18 fatty acids for developing high fat, high energy artificial "fruits" as bird food
- Control aggregation of lipoproteins and changes in phospholipids composition during extraction and storage of plasma
- Carry vitamin E and increase epidermal permeability of protective skin cream

- Create liposomes submerged in collagen gel to carry anti-inflammatory agents (AA) modifiers
- Establish phospholipid–phospholipid interactions to improve efficiency of degumming in oil processing

4.5 TROUBLE-SHOOTING OIL PROCESSING

A new seed plant (biological) has been developed by genetic engineering to produce food oils. It has high yields (high oil per seed), many seeds per plant and many plants per acre, but the oil is not very stable (i.e., it becomes rancid easily) and it has limited functional properties. It does not form a plastic fat when hydrogenated. The melting point is very low, and when solidification does occur, it is in the β form, although some sandy crystals form during storage at normal refrigeration temperatures.

You are a food oil refiner who has been called in to help before the project is thrown back to the plant geneticists. The oil has been refined, but perhaps not efficiently or appropriately. Propose explanations for the following, based on the information about the oil composition given in Table 4.2.

TABLE 4.2
Oil Composition

Oil Composition:		Fatty Acids in TAGs:	
		Arachidonic	3%
Triacylglycerols	99.8%	Linolenic	15
Tocopherols	0.01	Linoleic	55
Carotenoids	0.1	Oleic	8
Chlorophyll	0.05	Arachidic	3
Other	0.04	Stearic	15

(a) Why is the oil not stable? Give at least four reasons. Hint—think about the source of the oil. Suggest ways to overcome this problem.
(b) As an oil processor, what can you do to modify the oil to improve its functional properties and tailor the final products to specific product needs? Give at least two approaches. Be specific about what you will modify and why, and what effects this modification will have on the oil.
(c) What suggestions for modifications, if any, would you give to the plant geneticists for their next generation of genetic modifications of these plants?
(d) What information that you need to solve this problem are you missing?

4.5.1 Teaching Concepts

- Endogenous factors affecting oil stability—antioxidants, pro-oxidants, unsaturation

- Processing factors affecting stability—contaminants potentially introduced, pro-oxidants removed, heat, metals
- Evaluation of processing procedures and adequacy
- Identification of missing information
- Can refining "fix" a poor oil?
- Can finishing process "fix" poor refining?
- Tailoring winterization, hydrogenation, inter-esterification
    - selective rather than broad changes
    - achieving desired changes without creating additional problems
- Oil processing partnership with end user and research
- Numbers count (using quantities as clues in problem-solving)

4.6 USING FAT CONSTANTS TO IDENTIFY UNKNOWN OILS

The Calamity Oil Refinery ships a variety of oils worldwide. The U.S. Customs Bureau intercepted a shipment of Calamity Oil in which the outer cases were labeled olive oil but the inner bottles had no labels. The oils were tested and found to have the following characteristics in the standard oil tests:

IV	115–120
n_D	1.464 (15°C)
SV	194
mp	–2 to 2°C
sp grav	0.920

(a) From these values, determine the most likely identity of the oil. Remember, it could be pure oil, a modified oil or a mixture of oils. You will accept or reject the shipment for import based on your determinations.
(b) Give your reasoning for your assignment.
(c) What information that you need to solve this problem are you missing?

4.6.1 Teaching Concepts

- Use of fat constants to identify fats and oils
- Variation of fat constants with small changes in fatty acid composition
    - Saturation/unsaturation (# of double bonds)
    - *Trans* and positional isomers
    - Oxidation
    - Other adducts
- Potential effects of storage and other conditions on fat constants
- Develop sense of how much each constant varies with specific fatty acid changes
- Develop sense of which constant provides the best clues and information about oil identity
- Standards of identity for specific fats and oils
- Fat constant accuracy and sensitivity
- Regulatory and quality control use of fat constants—appropriate and inappropriate

4.7 INTEGRATING CHEMICAL AND INSTRUMENTAL ANALYSES TO DETERMINE LIPID COMPOSITIONS

You are a working cattle and dairy farmer who also does research for the Agricultural Experiment Station at Podiddle University. With all the negative publicity being given to beef and dairy products for being high in saturated fats, you decide to attempt to modify the fatty acid content of steaks and milk by feeding your animals diets high in fish oil or soybean oil. The feeding trials go well, your animals are still alive and healthy, and the milk doesn't smell fishy. However, now you are faced with the problem of determining how efficiently the fatty acids in the animal diet replaced endogenous fatty acids in tissues and in milk.

(a) Construct flow charts to show how you would isolate and separate the lipids of beef muscle and of milk.
   - What differences, if any, would there be between extraction methods used for the two lipid sources? Why are these necessary?
   - What differences do you expect in the lipid and fatty acid contents of the two tissues? Why?
   - How do these differences affect the isolation and separation procedures you will use?

(b) Describe
   - "Short ways" (polyunsaturated fatty acids [PUFAs] incorporated—yes or no).
   - "Long ways" (more complete and specific methods for fatty acid composition, localization and position) for determining the extent to which the PUFAs were incorporated into membrane and milk lipid structures. You don't have to list minute details of techniques; give the test or procedure you would use, general conditions, what the test(s) would tell and how the information would be provided.

(c) What information necessary to ensure success of these analyses are you missing? Where can you find the missing information?

4.7.1 Teaching Concepts

- Lipid composition/localization of muscle vs. milk
- Lipid extraction procedures:
  - Sample preparation and pretreatment
  - Solvents
  - Specific considerations for individual materials
  - Extract cleanup
  - Protection processes
- Methods for separating lipid classes—chromatography, chemical or physical fractionation
- Chemical/physical assays for lipid characterization—short chain fatty acids
  - (poly)unsaturation

differentiation of lipid classes and specific components

- Instrumental analysis of lipid components, distinguishing number and position of double bonds
- Positional analysis of TAGs and phospholipids

4.8 EMULSIONS

(This problem is adaptable to all levels, K–12 through graduate)

(1) What kind of an emulsion is mayonnaise? What are the phases? What is the emulsifier?
(2) What is the function of the vinegar, and why is it added to the egg yolks before the oil?
(3) Why can blender mayonnaise use more oil (1 cup oil) per egg than the mayonnaise made by hand ?

Explain the following specifications in terms of the chemical or physical processes involved in emulsion formation and stability:

(4) Have all the ingredients at room temperature.
(5) Beat the eggs yolks before adding anything else to them.
(6) Add oil slowly, in droplets, to the egg yolk initially.
(7) Use only 1/2 to 3/4 cup oil per egg yolk.
(8) Remedy a broken mayonnaise by beating some dry mustard into the mayonnaise in a warm bowl.
(9) Mayonnaise made in a blender must use the whole egg, not just the yolks.
(10) Explain the physical basis for each of the following remedies for broken mayonnaise emulsions suggested by other sources:
  (a) adding an additional egg yolk
  (b) chill the mayonnaise while beating

4.8.1 Teaching Concepts

- Critical roles of force and emulsifier in making emulsions
- Recognition of theoretical concepts at work in a real product
- Factors affecting emulsion formation and stability
  - Functions of individual ingredients
  - Effects of mixing method on emulsion stability
  - Effects of mixing force on particle size, stability, and potential for destabilization
  - Influence of oil-emulsifier and other ingredient proportions
  - Influence of order of addition
- Effects of various emulsifiers and stabilizers:
  - Hydrophobic-hydrophilic balance
  - Particles (wettable vs not wettable)
  - Polymers on oil surface

Multiple emulsifiers (e.g., in ice cream)

- Effects of aqueous phase viscosity }
  Size of oil droplets } Stokes Law Factors
  Oil:water ratio }
- Destabilization time/rate

4.9 LIPID DEGRADATION PROCESSES

You are the director of the analytical lab of an international oil processing company. The documentation of several tanks of oil was misplaced. Now your analytical crew must determine what the oil is and how it has been processed or treated because countries in the European Economic Community have very strict regulations about what they allow. In particular, EEC countries do not allow irradiated products, and U.S. regulations do not allow use or marketing of heat-abused lipid products. Also, health conscious food manufacturers are now setting more stringent requirements limiting acceptable levels of all forms of oxidation and other degradation of oils. You must determine what can be done with this oil. The principal TAG component of the oil is:

```
                  O
                  ||
          CH2-O-C(CH2)9CH=CH-CH2-CH=CH(CH2)6CH3          (R1)
   O       |
   ||
R2-C-O-CH
           |
          CH2-O-C-R3                          (R2 and R3 same as R1)
```

(a) Write full breakdown reactions that occur for this TAG when it is γ-irradiated, photosensitized, heated and(or) autoxidized. Include molecular formulas/structures, specific reaction pathways, names of major products expected for each degradation pathway.

(b) Identify the major products that are characteristic of each form of degradation. Which specific products or product classes would be most useful marker compounds for screening degradation mechanisms in processed fats and oils? How could each of these best be detected?

(c) What major changes in lipid properties or characteristics would be expected from each general degradation pathway? From specific products in the pathways?

(d) If Compound B below is found in this oil, can the oil be marketed in either the U.S. or Europe?

```
                                   H2C(CH2)4CH=CHCH2CH=CH(CH2)6CH3
                                    |
CH3(CH2)6CH=CHCH2CH=CH(CH2)6CH2
```

4.9.1 Teaching Concepts

- Practice in writing out degradation reactions and determining expected products for specific fatty acids or TAGs (as opposed to looking up products in tables)
- Reinforce degradation pathways and organic chemistry background.
- Comparison of degradation pathways and mechanisms:
  - Recognition of common products
  - Recognition of parallels between mechanisms
  - Distinction of products specific for each degradation
- Appreciation for the breadth of products resulting from degradations (connect to problems with analysis)
- Consequences of degradation—alteration of physical, chemical and sensory properties/quality
- Regulations and quality control issues

4.10 OXIDATION BY LIPOXYGENASE

Farmer Brown was an organic gardener who decided to test his farm produce for lipoxygenase (LPOx) to determine whether it would have great shelf stability. He made up a phosphate buffer solution at pH 7.5, added some soybean oil and analyzed LPOx contents in peas, tomatoes, grains, potatoes and cucumbers by measuring lipid hydroperoxides. He encountered problems with his assay, observing apparently very low activity for all of the enzymes except from tomatoes. He also found variable products, including some C10 and C12 hydroperoxides and derivative aldehydes, dihydroxy compounds and a variety of carbonyls. You are a consultant called in to evaluate what is going wrong and then "fix" Farmer Brown's assays.

(a) What is wrong with Farmer Brown's assays? Why?
(b) What information necessary to solve this problem is missing? Based on Farmer Brown's results, make assumptions about what you think he did. Explain how these procedures would affect assay results. Now use these procedures as a basis for recommending changes (see (c)).
(c) How would you modify the assays to make them accurately reflect lipoxygenase activity and reactions in each of the vegetables? Be specific, explain your reasoning in proposing the changes, as well as how and why you expect the assay results to change.

4.10.1 Teaching Concepts

- Lipoxygenase reaction mechanisms, expected primary and secondary products
- Identification of multiple dimensions or facets of "problem"
- Recognizing potential sources of degradation or inconsistency

- Factors affecting apparent lipoxygenase activity in vegetable extracts:
  - Sample preparation (does it make enzyme available?)
  - Plant source
  - Possibility of isozymes
  - Enzyme and substrate concentrations
  - pH
  - Fatty acid composition of plants
  - Oxygen concentration in solutions
  - Specific substrate used in assay
  - Potential inhibitors (e.g metal complexers)
  - Autoxidation confounding of LPOx reaction
  - Competition from hydroperoxidase activity of LPOx, from lyases
  - Assay method
- Factors affecting overall reaction of lipids in extracts
  - Sample preparation and handling
  - Contaminants (e.g., metal) in reagents, water, glassware
  - Buffer components (purity, metal complexing by PO_4)
  - Exposure to light
  - Photosensitizers in plant extracts

4.11 CONTROLLING OXIDATION

The Quality Food Corp. manufactures and markets dried soup mixes. They have found that their chicken noodle soup mix has a stable shelf life of only a few weeks. The mix consists of chunks of dehydrated chicken, dry noodles, salt and a chicken fat-based seasoning blend in dehydrated broth, packaged together in a clear cellophane package with colored printing. During storage, off-odors develop and are quite noticeable when the package is opened. Off-flavors develop in both the broth (after reconstitution in water) and the meat chunks. The meat chunks become tough; the seasoning blend and meat chunks darken. You are vice-president for research and have been called in by the board of directors to account for the problems. What information will you provide?

(a) What chemical reaction(s) and compounds are most likely to be causing the problems described for the

Odors
Flavors
Texture changes
Color changes

Be specific and show reactions.

(b) What information is missing that would help identify key reactions, products or problem spots? What could you do to find or generate that information?

(c) Suggest approaches to solve the problems and create a product that should retain its quality during storage for at least 6–12 months. Be creative; be specific. Explain the reasoning behind your suggestions. Discuss advantages and possible shortcomings of your proposed process.

4.11.1 Teaching Concepts

- Lipid oxidation in real foods vs. reactions on paper:
    - Direct products
    - Secondary products

  Consequences beyond lipid oxidation products (co-oxidation of proteins and other molecules)
- Recognition of pro- and antioxidant factors in complex systems:
  Fat composition
  Food matrix
  Sources of metals, hemes, photosensitizers, amino acids
  Sources of endogenous antioxidants
  Moisture content and water activity
  Light exposure
  Packaging permeability—light, moisture, O_2
- Looking "beyond the lamp post"—other molecules and reactions that may be active and interactive with lipid oxidation
  Co-oxidations of protein, starch, seasonings, pigments
  Maillard browning
  Packaging migration
  Phase partitioning of lipox or other products
- Tailoring effective processing and antioxidant "corrections" that are safe, appropriate, cost-effective

4.12 EDUCATION PERSPECTIVES ON TEACHING BY PROBLEM-SOLVING

Although the problem-solving approach described in this chapter evolved from personal experiences and perspectives and no education background, retrospective evaluation shows that it has solid grounding in educational and learning theory.

(1) It provides a virtual laboratory in which students personalize learning as they relate problems to their own experiences. This makes students more ready to learn and more receptive to new information and new approaches to thinking and reasoning.
(2) It creates a culture in which even mistakes or misunderstandings become opportunities for learning as students explore, debate, question and probe situations in depth.
(3) It facilitates teaching to different learning styles. Students differ in how they take in information (seeing it, hearing it, reading it, using it), whether they respond more instinctively to concrete or abstract content, and whether they process new material by active experimentation or reflective observation (Figure 4.4). Dealing with multiple learning styles is perhaps a professor's greatest challenge in reaching all students in the classroom. Using problem-solving in a variety of formats is exciting and rejuvenating on the teaching side as well as the learning side because it offers so many

Accommodators (learn by doing) *"What would happen if I did this?"* Input: concrete examples Processing: active experimentation Thinking mode: concrete-active (activist) Teaching modes: independent discovery	**Divergers** (observe concrete then think) *"Why does this happen?"* Input: concrete examples Processing: reflective observation Thinking mode: concrete-reflective (reflector) Teaching modes: lecture with analysis and application, thinking time after questions
Convergers (listen to abstract - think-do) *"How does this happen?"* Input: abstract concepts Processing: active experimentation Thinking mode: abstract-active (pragmatist) Teaching modes: Interactive, problems, assignments requiring active involvement and applications of details	**Assimilators** (think abstract–observe concrete) *"What is there to know?"* Input: abstract concepts Processing: reflective observation Thinking mode: abstract-reflective (theorist) Teaching modes: authoritative lectures with demonstrations, lab exercises with detailed tutorial

FIGURE 4.4 Patterns of learning, based on differences in how students take in and process information (work of Kolb as described by Little).

options and opportunities for dealing with information. Somewhere in the mix, each student finds a path that fits.[5,10]

(4) It integrates all aspects and stages of learning. Learning involves conditioning, reinforcement, memory process, concept formation, problem-solving and creativity in four phases:[11]
   - Receiving information—lectures, class notes, references, Internet, discussions
   - Processing information—thinking, evaluating, applying, questioning, integrating with prior knowledge or experience, memorizing
   - Storage or retention—encoding information for retaining in memory
   - Retrieval or recall—recognizing and recalling stored information appropriate for use or application

   Long-term and lifelong learning strategies are strengthened when connections are continually made between information delivered in class (Stage 1), thinking processes (Stage 2), understanding information logically so it is retained without "memorization" (Stage 3) and recognizing already known facts that are relevant to the issue (Stage 4).[11] Pulling this all together at once has a very powerful effect on students—it definitely changes the way they think.

(5) It transforms thinking and learning. Problem-solving transforms thinking through cognitive reframing, training people to think differently about difficulties that cannot be changed and to think more positively about tough subjects.[12] Rather than being overwhelmed with the details of the physical

and organic chemistry of lipids, students see first-hand how the chemistry drives real applications, and this makes it easier to connect and understand. Working on real-life problems removes mental stumbling blocks against learning the chemistry and gives students a reason to learn. Once initial associations are established, students become more familiar with the underlying chemistry and more willing to use it. They learn how to concentrate on relevant and important points rather than irrelevant ones. This total process provides a strong positive reinforcement to motivate students.

Problem-solving transforms learning by discarding old habits of dealing with information, opening up frames of reference, forcing consideration of alternatives, and changing perspectives.[5,6,13–16] Real-life situations in problems provide effective and realistic focus for reflection; they lead students to recognize questions beyond the obvious and superficial, probe unspoken assumptions, explore alternatives, analyze consequences of choices and actions, and thereby make meaning of their world. The process required to arrive at solutions is a rational process that dynamically and actively links theory to practical behaviors or properties and thus forces students to reevaluate how they see, think about and understand the world around them. The end result is a transformation in thinking that becomes the foundation of lifelong active, thoughtful learning.

To close, I will admit that teaching by problem-solving is very intense. For the professor, it is more work than lecturing because there is never a time to sit back and slide on routine. There are problem sets to grade every week, new applications to find and a kind of constant internal pressure to be creative and interesting. The positive counterbalance is that problem-solving is much more stimulating and fun, and working with each new group of students, even on the same problem, brings different challenges, ideas and perspectives. Problem-solving keeps teaching energized and fresh.

For students, it has been said that this course is not for the faint of heart. There are no weeks for sloughing off between exams and the intense higher-order thinking required can sometimes scare and overwhelm students initially. However, any advanced course that covers a lot of material in depth is daunting to students. Problem-solving actually overcomes this by dealing with small amounts of new material at a time and integrating it with previous learning and with real life before moving on. In addition, dividing class sessions between modified lectures and problem-solving discussions provides both traditional and innovative learning forums for students.

What do the students say about learning by problem-solving?

- Students universally report that learning through logical development and through connections of fundamental properties and reactions to real-life situations is more effective than memorizing massive amounts of material to regurgitate factually on exams.
- Students find that using hands-on examples and working with problems helps them conceptualize new information and connect it to things they

already know. As a result, they understand concepts better, remember them long-term and are able to use and apply them practically. They also learn to recognize when knowing and applying details is important.

- Through problem-solving, students learn large amounts of material "painlessly," without memorizing and without realizing how much they have retained. When material is constantly reinforced throughout the semester, students find they can answer most of the questions on a "working knowledge" final exam without studying or looking up the answers.
- By comparison, when traditional exams were added back to the course, students felt overwhelmed by material and focused all their mental efforts on memorizing to answer questions precisely rather than thinking how to use information and letting the connection and retention follow. Students have repeatedly elected weekly problems over term papers and exams.
- Examples and problem-solving make classes more interesting and challenging. Several sets of part-time students from industry have even taken the problems back to their companies to use as exercises in group meetings.
- Problems and discussions guide students beyond the printed page and obvious answers, and show them how to recognize multiple aspects of a question and to evaluate multiple solutions.
- Learning skills developed through problem-solving can be applied to other classes and all aspects of life.
- Students develop self-confidence and a sense of achievement far beyond traditional lecture–exam approaches.
- Problems make learning FUN!!!

Quotes from students:

"This course has completely changed the way I think and deal with problems at work."

"This approach will benefit my whole life and career. It is excellent, interesting and challenging."

REFERENCES

1. Cambourne, B., Conditions of literacy learning. The conditions of learning: Is learning natural?, *The Reading Teacher*, 55 (8), 758, 2002.
2. Illeris, K., Towards a contemporary and comprehensive theory of learning, *Int. J. Lifelong Educ.*, 22 (4), 396, 2003.
3. Boisen, L. and Syers, M., The integrative case analysis model for linking theory and practice, *J. Social Work Educ.*, 40 (2), 205, 2004.
4. Cook, N., The age of discovery, *Instructor*, (March), 23, 2006.
5. Cranton, P., Individual differences and transformative learning. In *Learning as Transformation*, Mezirow, J., ed.; Jossey-Bass, San Francisco, 1991, 181.
6. Cranton, P., Teaching for transformation, *New Dir. for Adult and Continuing Educ.*, 93 (Spring), 63, 2002.
7. Cranton, P. and King, K. P., Transformative learning as a professional development goal, *New Dir. for Adult and Continuing Educ.*, 98 (Summer), 31, 2003.

8. van Merrienboër, J. J. G., Kirschner, P. A., and Kester, L., Taking a load off a learner's mind: Instructional design for complex learning, *Educ. Psychologist*, 38 (1), 5, 2003.
9. Shepard, L. A., Linking formative assessment to scaffolding, *Educ. Leadership*, (November), 66, 2005.
10. Little, L., Lessons in leadership: Kolb's learning styles for leaders, *Administrator*, (August), 8, 2004.
11. Alutu, A. N. G., The guidance role of the instructor in the teaching and learning process, *J. Instruct. Psychol.*, 33 (1), 44, 2006.
12. Chang, K., Teaching a tough subject from a perspective of psychology, *Math. School*, (September), 32, 2005.
13. Mezirow, J., Perspective transformation, *Adult Educ.*, 28, 100, 1978.
14. Mezirow, J., *Transformative Dimensions of Adult Learning,* Jossey-Bass, San Francisco, 1991, 145.
15. Mezirow, J., In *Learning as Transformation: Critical Perspectives on a Theory in Progress*, Jossey-Bass, San Francisco, 2000.
16. Kasl, E. and Eliad, D., Creating new habits of mind in small groups. In *Learning as Transformation: Critical Perspectives on a Theory in Progress*, Mezirow, J., Jossey-Bass, San Francisco, 2000, 229.

5 Mentoring Independent Study Students in Lipid Science

*E. Chris Kazala, Crystal L. Snyder, Tara Furukawa-Stoffer, Randall J. Weselake**

CONTENTS

5.1 INTRODUCTION

Most researchers in the area of lipid science are familiar with the desire to develop value-added consumer products. By improving a basic ingredient or a processing step, value can be added to all downstream products, resulting in some societal or monetary benefit. By analogy, exposing undergraduates to research early in their careers provides an opportunity to add value to their science education, resulting in students that are better equipped for any number of career paths in science. Undergraduate research opportunities forge a vital link between theoretical knowledge and practical application that is often absent in theoretically biased undergraduate programs. Many colleges and universities acknowledge the value that exposure to a

*Corresponding author

practical research project may provide, encouraging students to fortify their degree with research projects and sometimes even allowing for special degree designations (e.g., cooperative education degrees) if certain work-term requirements are met.

The following is intended to serve as a guide for the researcher who relies on contributions from undergraduate students in a research laboratory setting. Although teaching researchers (i.e., professors) are the most likely audience, the considerations and guidelines set forth may be of interest to others who are involved in such research projects. These will include scientists from public institutions and those from industry, as practical and cooperative education work terms are not limited to post-secondary institutions.

Over the past 15 years, the authors have directly supervised or co-supervised over 190 undergraduate student projects, mainly in the corresponding author's previous lab at the University of Lethbridge, Alberta. The projects took the form of "Independent Studies" (credit course, minimum time commitment of 10–12 h/wk), cooperative education work terms (credit course, full-time paid positions) and summer student work terms (full-time paid positions, credit obtained with an optional evaluation component). All projects were 4 months in duration, although many students went on to work in the lab for multiple terms, often continuing work on the same general project. Several continued on to successfully complete graduate degrees within the lab, while others have gone on to other labs and professions. A summary of the present whereabouts of those we were able to keep track of is shown in Figure 5.1.

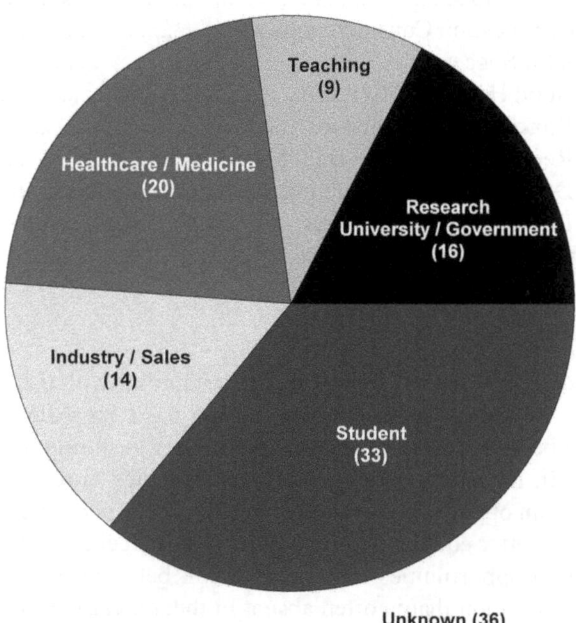

FIGURE 5.1 Career paths of 92 former students who have completed undergraduate research projects in our lab over the course of 16 years. A total of 128 students (67 female, 61 male) completed 194 individual projects, averaging 1.5 projects per student.

It is our intent to provide some general and specific considerations to the teaching researcher involved in the delivery and supervision of these types of projects. Topics that will be addressed include project development, student recruitment, supervisory tips, student evaluation and authorship issues. Although the authors' collective experience is in the supervision of student projects in a lipid biochemistry environment, many of the principles can be applied to most research settings where undergraduate students are part of the laboratory.

5.2 RECRUITMENT—FINDING THE RIGHT STUDENTS

Many students at this level of their training are keen to undertake research projects, regardless of their future career aspirations. For some, an undergraduate research project might be viewed as a chance to bolster an application to professional or graduate school, for others an opportunity for summer employment and some may undertake research simply to explore their options and interests. In any event, it is important to consider more than their academic performance or career ambitions when recruiting students for independent studies. Unless scholarship eligibility (e.g., summer studentships) is of the utmost importance, the student's attitude is perhaps the most important attribute to consider. Often the most academically gifted students do not possess the right attitude toward research and may, in fact, be detrimental to overall lab morale. It has been our experience that many students with slightly inferior transcripts can excel outside of the structured classroom environment and, if given the opportunity, will often thrive in a research lab with adequate mentorship.

Attitude will also, to a large extent, determine how well a student will fit within the research group, a factor that should not be overlooked if the student is to work closely with other students and technical staff or perhaps even serve effectively as an ambassador between collaborators (e.g., if co-supervised). While overall attitude may be difficult to assess and predict, a brief interview process will provide the principal investigator (PI) and other supervisory staff with an idea of the student's personality and motivation for undertaking a research project. A curriculum vitae (CV) or resumé can also be revealing, particularly if it contains obvious mistakes that provide insight into an applicant's attention to detail. Previous work or lab references and the PI's own level of experience in applicant evaluation will also prove to be valuable in helping to choose suitable students.

Teaching senior-level (i.e., third- and fourth-year) undergraduate courses provides an excellent opportunity for recruiting students for undergraduate research projects. The length of semester courses gives the teaching researcher ample time and opportunity to evaluate potential students based on their academic performance, class participation and overall attitude. The opportunity to interact with students for several months offers researchers a preview of how a student may fit within their particular research program. To complement this situation, the course curriculum at this level is sufficiently advanced to permit the incorporation of relevant research material in the classroom, giving students an opportunity to develop an interest in a particular research area. Students are often introduced to the instructor's research program through in-class discussions, the assignment of term papers or presentations related to the instructor's area of research, and identification of current gaps or limitations in the area of research that need to be addressed.

Maintaining an open dialogue with teaching colleagues is also an excellent way to discover and evaluate potential students. Students who are referred from other faculty members, or whose attributes may otherwise be verified from trusted sources, remove some of the guesswork from the selection process.

Developing and maintaining a lab website will attract students from other institutions. While they represent a low percentage of all the students who have passed through our lab, they have made meaningful contributions to our efforts. Out-of-town students tend to seek full-time work terms, often during the summer when they are returning to visit family for the summer break. Our web page (http://www.afns.ualberta.ca/hosted/wrg/) has also been instrumental in attracting postdoctoral fellows (PDFs) and graduate students as well as serving as a store of information concerning manuscripts, course materials and assignments, and contact information for lab members.

5.3 PROJECT DEVELOPMENT

The primary differences between undergraduate- and graduate-student projects are time and scope. While graduate projects are intended to span several years and typically begin somewhat open-ended in scope, undergraduate projects need to be manageable over a short term (4–8 months is typical). The challenge when developing a project for an undergraduate student is to keep it manageable within the limited time frame and substantial enough to maintain the student's interest and encourage an element of ownership over the project. This latter condition cannot be overemphasized. Too often where undergraduate projects fail, it is because students lose interest or become just cogs in an assembly line operation where their contribution to the bigger picture is either ill-defined or perceived as inconsequential. Despite the common perception, undergraduates are not merely cheap technical help. While their contributions may relieve graduate students or technical staff of relatively simple, tedious or repetitive tasks, a learning experience is expected and must be delivered, particularly where students pay tuition for the opportunity. Students who are denied an adequate learning experience typically lack incentive to take ownership of their project and will rarely excel in their role. When developing a research project, it is almost always better to carve out a well-defined subsection of a larger project that they can realistically hope to see to completion rather than burdening them with repetitive tasks on large numbers of samples.

Unlike a graduate project, where students can be presented with a general research goal and be expected to take a role in developing their own research strategy, undergraduate projects require more detailed planning on the part of the supervisor. A flow chart that outlines the sequence of the main elements in the research project is shown in Figure 5.2. Several items to address in developing an independent study course outline follow.

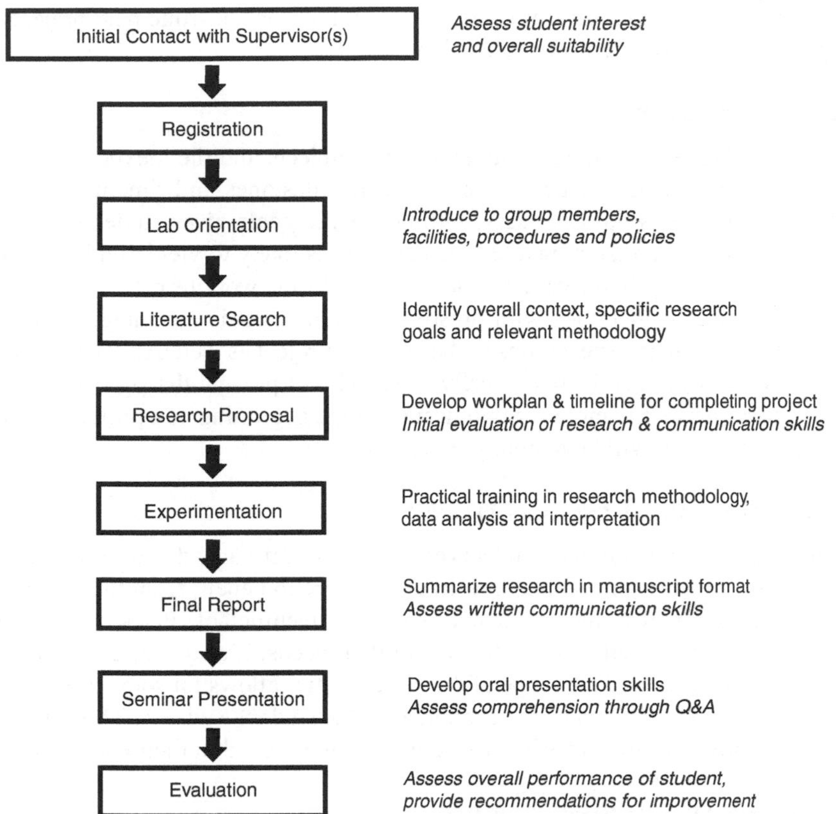

FIGURE 5.2 The progression of a typical student through an undergraduate research project. Notes on the right indicate the objectives of the student (regular text) and the supervisor (italics) at each stage.

5.3.1 The Research Objective

As with any experimental strategy, a clear understanding of the objective is required. Providing sufficient background information to put the project in its appropriate context will help the student better understand his or her role. For example, if students are to analyze food samples for *trans*-fatty acid content, they should understand the importance of this information in the context of the overall research efforts. If possible, background literature and review articles should be provided well in advance, perhaps as early as the initial meeting. A good start to a successful project often hinges on the students' understanding of what they are to do, and why it is important. We recommend that students conduct their own initial literature review and develop a short research proposal before beginning any work at the bench. While this ensures that they come into the lab with a basic understanding of their objectives, we have found that, with short-term projects and inexperienced students, considerable momentum might also be lost in the early stages of a project if too much time is spent on a written proposal. Currently, we feel it is more realistic, given the time constraints on such projects, to provide new students with substantial guidance with this requirement. Returning students and those on full-time paid projects are expected to develop their own research proposals.

5.3.2 Expectations

Given the short-term nature of undergraduate projects and the inexperience of the students themselves, it is best to set out clear milestones and timelines. While a concrete timeline is not necessary, setting specific goals gives students direction. It is our experience that extra attention paid here is likely to yield dividends downstream, particularly in the case of students on full-time work terms.

During initial meetings with prospective students it is advantageous to have them articulate their expectations of the research project as well. An open and frank discussion of what can be realistically expected will prevent disappointment later. In a general sense, the more communication at the beginning of a project, the more likely the outcome will be mutually satisfying.

5.3.3 Skills and Techniques Learned

Providing a list of techniques students can expect to learn ahead of time will provide some basis for them to evaluate how likely they are to benefit from the experience. It is imperative that immediate supervisors (e.g., technicians, PDFs, graduate students) also be informed, in order to plan for their needs. If a student desires to learn specific techniques and the scope of the PI's activities allows for some flexibility, an honest effort should be made to accommodate the student's wishes. This will result in more motivated students who are more likely to succeed in their endeavors.

5.3.4 Supervisory Responsibilities

Although the responsibility to evaluate the student typically lies with the PI, day-to-day supervision of undergraduate students often falls to other research staff or graduate students. Ensure that the responsibilities and expectations of the supervising team members are clear, particularly in the case of graduate students who may have limited supervisory or teaching experience. If two or more people will share the supervisory responsibilities, it is vital to ensure that only one is responsible for assigning duties and scheduling, to avoid unintentional conflicts. It is ultimately the responsibility of the professors to monitor their labs for potential interpersonal conflicts and to ensure that junior supervisory personnel conduct themselves appropriately and professionally.

5.3.5 Time Commitment

One should be realistic about the time commitment expected of the student. A minimum commitment of around 10–12 hours per week is typical for part-time undergraduate projects. Some students will attempt to carry a full course load in addition to their research project and must schedule their research into an already bursting schedule. If a certain procedure requires several consecutive hours to complete, and students are available for only an hour or two per day between lectures, how productive will they be? It is important to be clear with students about the time commitment required during the recruitment process and develop projects that can be somewhat flexible. Other members of the lab can be relied on to assist in this regard as well, provided this does not cause their own projects to suffer. Lengthy experiments or procedures can either be started or completed by full-time lab members, with the undergraduate student participating as time permits. If these situations arise frequently, a conscious effort should be made to expose the student to all aspects of the procedure, as opposed to coming in and performing the same task time and time again. It is important, however, to ensure that any assistance offered from full-time lab members is not abused, and to emphasize that the student is responsible for completion of the project. Full-time lab members should not be expected to "carry" a substantial portion of an undergraduate project due to the student's time constraints. Where a suitable compromise cannot be agreed upon in advance, students should be encouraged to postpone their project until their academic schedule permits adequate time commitment to the research.

We have also encountered a recurring situation where students involved in a full-time work term (e.g., to fulfill a cooperative education designation requirement) wish to enroll in another course or two during the semester. It has been our policy that co-op students seek the permission of the PI before enrolling in any course scheduled during the workday, with approval granted on a case-by-case basis. If approved, they are to document their time spent away from the lab as well as the extra time they have worked to make up for it. Time spent away from the lab for course work should be restricted to scheduled lecture or lab periods only. Additional study and completion of assignments should be done on the student's own time and should not become an excuse for failing to complete experiments. No full-time work-term student should be allowed to take more than one course during the regular workday. The make-up time required in the lab becomes too demanding to maintain, and the lab project invariably suffers. Students wishing to take more than one course should be encouraged to consider alternatives (e.g., carry a full course load for the term instead, and return for a summer work term). When considering allowing a student to work after regular hours in the lab, the institution's policies must be kept in mind and adhered to (e.g., working alone and safety policies).

5.3.6 Evaluation

All course outlines should address student evaluation. In a research project, student evaluation will often take the form of a final written report, oral presentation or both. It may also include other criteria, such as the quality of their bench work, record-keeping, ability to follow directions, professionalism, participation and attitude. The relative weighting of each will likely vary from lab to lab, as can the general expectations. These are important considerations for undergraduate students as they may work in several different labs over their course of study and arrive in a lab with preconceived notions of what the expectations will be. When possible, it is best to communicate to a student the types of actions that will result in excellent and substandard grades. We have found it useful to provide students with blank copies of evaluation sheets that their supervisor(s) will use when evaluating their performance. Figures 5.3 and 5.4 are examples of evaluation forms for oral presentations and final reports, respectively, that we have used in the past.

Presentation Evaluation

Student:________________________________ Date:________________________

Title:____

Please use the space provided for specific comments.

Organization (15) – Introduction, logical flow of ideas, conclusion or summary

---

Presentation style (15) – Speaking voice and pace, any distracting tendencies, reading slides instead of using them as a guide, appropriate use of humor

---

Content (40) – Appropriateness of material to a 2nd,3rd,4th year (circle one) course, explanation of concepts, overall knowledge of topic, ability to answer questions

---

Use of visual aids (20) – Legibility of slides. Includes background and font color/style/size, clarity of figures and slides, proper referencing and format (e.g., legends, axis labels)

Timing (10) – Deviation of more than 10% (above or below) the prescribed length will result in deductions

---

TOTAL /100 Evaluator initials:____________

FIGURE 5.3 Evaluation form used for the student's oral presentation of his or her research project. All senior lab personnel (graduate students, technicians, PDFs) should be encouraged to evaluate all students and submit completed forms to the PI for compilation and final evaluation. If resources permit, comments can be combined and typed for anonymity prior to providing the student with constructive feedback.

Term Paper Evaluation

Student:______________________________ Date:______________________________

Title:____

Please use the space provided for specific comments.

Content (60)

Organization (20)

Grammar, Style, Spelling (10)

Referencing (10)

TOTAL /100 Evaluator initials:__________

FIGURE 5.4 Evaluation form used for the student's term paper. As with the evaluation of the oral presentation, all senior lab personnel (graduate students, technicians, PDFs) should be encouraged to evaluate all students and submit completed forms to the PI for compilation and final evaluation. If resources permit, comments can be combined and typed for anonymity prior to providing the student with constructive feedback.

5.3.7 Certification Considerations

While developing a particular project, a supervisor should keep in mind the issue of safety and adhere to all governmental and institutional safety requirements. Many institutions require WHMIS (i.e., workplace hazardous material) training, and some of the more challenging projects may require radionuclide worker or biohazard certification. It is best that these issues be addressed before committing to a student project, because delays while waiting to attend a safety course may be considerable, allowing for very little bench work within a 4-month work term. Existing safety certifications are among the reasons to encourage good students to continue on in the lab for two or more projects.

5.4 STRATEGIES FOR SUCCESS

5.4.1 Good Habits and Skills

The teaching, development and continual reinforcement of good laboratory habits should remain foremost in one's mind when training and supervising an undergraduate student. As in life, bad habits forged early become more difficult to correct with age or experience. Therefore, regardless of what the student's past experience may be, it is best to start with the basics. This will not only help students build confidence and familiarity in their new environment, but will give the supervisor an opportunity to observe their technique and make corrections as required. Though it is sometimes assumed that teaching basic techniques is the responsibility of the teaching labs, taking the time early to ensure that a student's previous training meets the expected standard will give you both confidence in the data that is generated. If several students arrive in the lab with a similar problem, as evidenced by bad habits, it may be indicative of poor technique being taught in the teaching labs and should be discussed among faculty and lab instructors so that the error is corrected.

In almost any research lab, several simple procedures can be used to assess a new student's skill set, attention to detail and manual dexterity. In our lab, many projects involve the study of the lipid biosynthetic enzymes. Buffer preparation and colorimetric protein quantification assays often provide new students with the first experimental work of their project. We will often start by giving students a buffer composition and ask them to come up with a recipe for a specific volume of buffer (e.g., 250 ml), with written step-by-step instructions on how to prepare it. They are allowed to ask questions at any time, and often they do. Having them put their procedure to paper initially allows the supervisor to check their calculations, written communication style and even their attention to detail before they even touch a beaker. Once the recipe has been approved, students proceed to make the buffer under direct supervision, perhaps with a brief interruption to address the correct care and operation of the pH meter. It is an effective exercise in that they will learn where things are located in the lab (reagents, glassware, instruments) one step at a time and at their own pace. It will provide the supervisor with many opportunities for lab-specific rules (e.g., cleaning up the balance area, returning reagents to their shelves), and it will give the supervisor an opportunity to determine whether students will require more or less supervision than average during their tenure in the lab.

Protein assays based on colorimetry can also prove to be an effective exercise for both student and supervisor, even if a bovine serum albumin solution of unknown concentration is used in place of a sample. It is simple and rapid to perform. In this case, it is appropriate for the supervisor to carry out the procedure in front of students, and then have students repeat what was just shown to them. Replicates will provide student and supervisor with instant feedback on the reproducibility of pipetting technique. A dilution series of a protein standard solution, prepared by the supervisor but assayed by both, can be used for evaluation and to plot a standard curve.

Safety issues should also be addressed before or during this initial lab session. In our lab, in keeping with departmental rules, students are given a handbook of safety procedures to study before coming into the lab for on-site safety orientation. The location of safety equipment (i.e., spill kits, safety showers, eyewash stations, fire extinguishers) is pointed out and demonstrated where appropriate. Following this on-site training, students are expected to complete an on-line safety quiz, which acknowledges their understanding and acceptance of due diligence policy. Safe lab practices are enforced by all senior lab personnel and should be monitored by all lab workers.

5.4.2 Protocols and Procedures

Clear instructions and written protocols form the foundation of the undergraduate student's independent research project. While students are expected to take their own notes and record the protocol in their own words in their lab book, a standard written protocol that they can refer to will ensure that no part of the procedure has been altered by means of omission, miscommunication or misinterpretation. This establishes continuity and consistency that are essential, particularly since undergraduate projects under the same umbrella can change hands many times over the course of years. On a related note, care should be taken to annotate outdated protocols as such and make sure they are removed from circulation. We have had several instances where a student will attempt to use a procedure that has since been modified, causing future problems with interpretation and reporting. We recommend that, once a procedure or protocol has been modified, all loose hard copies (i.e., those not in a lab book) of the procedure be destroyed and that only electronic records be kept. If necessary, old copies may be removed to a private office. Outdated methods published in journal articles can prove problematic. It is usually the responsibility of the immediate supervisor to ensure that the proper protocols are being employed.

That being said, a protocol, no matter how clearly written, is no substitute for thorough hands-on instruction by qualified personnel. The protocol itself provides only the practical framework for a procedure and typically offers little in the way of theory or rationale. It is absolutely essential for supervisors of undergraduate students to take the time to rationalize each step of the protocols during the training process. Knowing what to do is very different from understanding why it should be done, or perhaps more importantly, why it should be done a certain way. This is also a key element of their learning experience, since it will help them gain a practical understanding of how methods are developed and how their work relates to general scientific method. For students to develop into independent critical thinkers, it is important for them to understand the rationale and underlying science behind the protocols.

Fortunately, the laboratory is a hands-on environment, where a demonstration speaks volumes. Capture the opportunity wherever possible to illustrate the underlying science in a hands-on, memorable way. For example, one mistake students often make with lipid extractions is to assume that the upper phase is always organic. An effective means of breaking this perception is to have them hoist a bottle of chloroform. Often, even the simplest demonstration can enforce a fundamental concept.

Do not be afraid of engaging and questioning students. Ask questions that force them to think beyond the protocol. Encourage them to use their theoretical knowledge to rationalize their methods or results themselves. Challenge them to defend their explanations by posing hypothetical alternatives. In the lab it is often easy to get caught up in protocols and forget that, at its core, science is a very powerful approach to problem solving. It is just as important to help students develop this way of thinking as it is to teach them how to use the tools and techniques of the laboratory.

Finally, remember that even with the best training, mistakes happen and technical issues will arise. It is important, particularly with undergraduates, to keep mistakes in perspective and resist overreaction so as not to needlessly shatter a student's confidence. Approached tactfully, mistakes can often be turned into valuable teaching opportunities.

While this approach requires a certain amount of time and patience that may not seem worth it for such short-term projects, the value of the learning experience for the student increases dramatically and students are more likely to return for additional semesters, summer work terms, or even as graduate students. In the shorter term, students who are thoroughly trained and encouraged to think independently tend to have greater confidence and are better able to address any difficulties they may encounter. As they gain experience, they are more likely to assume a leadership role and are often able to contribute as effectively as other, more academically qualified members of the research team.

5.4.3 Record Keeping and Data Management

Maintaining a laboratory notebook is one of the most important, if often underemphasized, skills that students must learn over the course of their research project. With short-term undergraduate projects that may be passed from student to student a number of times before publication, the emphasis on adequate lab notes cannot be overstated.

Developing a standard set of guidelines for laboratory notebooks, and in some cases, even sample labeling and storage, is the simplest means of establishing a common expectation for what is required of students. These should be explained thoroughly prior to any research being undertaken, and it is ideal to have the supervising team member read and offer feedback after the first several entries are made, to correct any deficiencies early. This will help ensure that key details are not missed that could compromise the publication of later experimental results.

It is also worth emphasizing the need to track all experimental samples from "cradle-to-grave" again, to best facilitate the passage of project materials between students. Whether there is a lab-wide coding system in effect for samples of a certain type or not, students should be reminded to clearly identify their samples in storage and in their lab notebooks. In our lab, it is customary for departing students to record an inventory of any research materials they have in storage on the last page of their notebook. Near the end of the project, a walk-through with a more permanent member of the lab, detailing the physical location of important files and samples, is encouraged.

Data in electronic formats should, where possible, be included in the lab book as hard copy; however, for publication purposes at a later date, it is usually preferable to have access to an electronic version of their data files and any written work associated with the project. Students should be required to provide a disc backup of all of their files that can be located with their lab notebook for future reference. Filenames of the associated data files may be included in the relevant lab book entries.

Some students may prefer to type their lab book entries and paste them into their notebooks. While this does carry certain advantages, such as increased legibility, standardization of formatting and the availability of an electronic copy with integrated data files for searching at a later date, there is a major caveat to this approach. There is a risk that important details may be missed or forgotten if students are not forced to take good notes while working at the bench, because there is often a tendency for them to wait to record their notes until they can get to a computer to type them. The easiest way to combat this is, of course, to have them complete their notebook entries as they work. If a computer-generated version is to be completed, lab notes should still be recorded at the bench in a separate notebook. Students should also be reminded to paste hard copies of their entries into their book on a daily basis, to prevent accidental and potentially catastrophic data loss in the event of a computer malfunction.

5.4.4 Developing and Evaluating Written Communication Skills

Most undergraduate projects include a written component in the evaluation, usually in the form of a final research report or review. Writing is a common weakness among undergraduate science students, as scientific accuracy is valued over clarity of composition and appearance in many undergraduate classes. This is another area, however, where establishing good habits will yield considerable benefits over the long term.

Students should be encouraged to prepare publication-quality reports free of careless grammatical, typographical or formatting errors. Carelessness in written work should not be condoned any more than carelessness at the bench, particularly because such mistakes have the potential to reflect poorly on the value of the experimental data itself.

The appropriate use of references should be included in the evaluation criteria for written work. Too often, there is a heavy reliance on review articles or non-peer-reviewed material, for example, from Internet sources. Such inappropriate citations should be strongly discouraged. It is common practice in our lab to limit the use of review articles to two and also to institute guidelines on how recent the chosen references should be (i.e., most references should be from the last 5 years). Incorporation of figures and tables from published references should be discouraged in favor of synthesizing original figures based on information from multiple sources. Students should be instructed on the appropriate citation formats or the use of citation-management software. Many academic libraries offer discipline-specific workshops or publish tip sheets on the appropriate use and formatting of citations for students. If no such program exists at your institution, science librarians or reference specialists are usually available on an appointment basis for consultation. Some institutions even offer one-on-one writing consultation services that may be beneficial to students who require assistance with editing or proofreading. These services greatly relieve the time burden on technical staff who may otherwise be called upon to assist students in this area, and students in need of assistance should be encouraged to avail themselves of these services where they are offered.

Good time management will also in many cases drastically improve the quality of written work submitted. While there is often a temptation to put off writing until near the end of the semester, the literature review may yield important insights into experimental results, which may not be explored if uncovered too late. This is particularly frustrating in situations where students have struggled with technical difficulties early in their projects, only to find insight at the end while preparing their report. Requiring students to submit an abstract near the midpoint of the semester will encourage them to get started on their literature reviews at an earlier stage.

Although publication of undergraduate research papers has not traditionally been commonplace, and many supervisors prefer to pursue more conventional avenues of publication, over the past several years there has been a steady rise in the number of publications catering specifically to undergraduate research. While some of these publications are discipline- or even institution-specific, many publish across disciplines, and some have formed undergraduate review boards to expose students to both ends of the publication experience.

5.4.5 Interpersonal Challenges

Occasionally, problems will arise in spite of all the care and planning that goes into student selection and project development. Any number of challenges or conflicts may occur that can affect student performance or overall lab morale, and it falls upon the shoulders of supervisory personnel to resolve such issues before they become unmanageable. In such circumstances, it is imperative that those involved in conflict resolution maintain an appropriate level of professionalism and personal objectivity. Common sense and a cool head will generally be much more effective at addressing any problems than temperamental outbursts. Remember that attitude reflects leadership, and that not all problems start with the student.

Motivational issues are one of the most common problems that arise in undergraduate projects and can be one of the most challenging to address. When students do not perform the tasks set forth, show a general disinterest in their project or only do just "enough to get by," it is tempting to blame the students for being lazy. While it is sometimes the case that students are lazy or some other personal circumstance is preventing them from performing as expected, a lack of motivation can sometimes also be a symptom of poor leadership or project planning. Undergraduates, by virtue of their lack of experience, are particularly susceptible to slipping into idleness if they become bored or frustrated with some aspect of their work. It is therefore important to maintain open communication between supervisor and student to ensure that a student is making progress and that any problems are addressed promptly to avoid losing momentum. The principle of inertia applies to students doing research as well as it does to physics. It is much more difficult to get students moving again once they have stalled than it is to keep them moving in the first place.

Students possessing a lack of confidence are another problem frequently faced during their initial foray into the research lab. This is to be expected to some degree from all students who lack experience in the lab and can often be overcome once the first few experiments are in hand. For supervisors, the problem lies with those students who continue to need continuous "handholding" even after they have become oriented with their protocols. Sensitivity and patience are required here to avoid compounding students' situation further with the feeling that they are being disciplined. Often, such students will ask questions to which they know the answers just because they lack confidence in their own knowledge. Rather than responding with frustration or annoyance (even if it is the tenth time they have asked the same question), or giving them the answer outright, a good strategy is to ask them what they think the answer is, or provide some direction as to where they might find the answer for themselves. Often they will come up with the right answer on their own. Even a simple rephrasing of the problem in different terms may be enough to trigger them to work it out on their own. If mistakes are made, keep the criticism constructive. If there are certain techniques or procedures they seem particularly uncomfortable with, ensure that they have received adequate training on the protocol and the underlying rationale.

Personality conflicts among students and other lab members are relatively uncommon, but can be particularly disruptive due to the nature of common work areas and the tendency for people to take sides. While circumstances dictate that these types of conflicts be dealt with on a case-by-case basis, the supervisor must be aware of the effect such conflicts have on overall lab morale and take swift action to prevent divisions from forming among other lab members.

A special case worth noting is that of a conflict's arising between a supervisor and a student. Such conflicts must be addressed tactfully to avoid further problems from developing. In situations where the student may answer to more than one supervisor, care must be taken that one does not unintentionally undermine or contradict the other. Maintaining open communication between supervisory personnel is vital, and complex situations may require intervention or corrective action undertaken by the ultimate authority in the lab, the PI.

As in most situations where human beings interact with each other over prolonged periods, romantic relationships between students and other lab personnel occasionally develop during the course of a project. While it is difficult to discourage or control relationships between students, relationships between students and supervisory personnel must be avoided, and supervisors should be informed beforehand of any existing institutional policy in this respect. Even so, human behavior can be difficult to predict and such developments should be dealt with promptly. Most modern institutions recognize that personal relationships may indeed develop and have developed policies to address such scenarios. If a draconian edict simply forbidding relationships is put into effect, one runs the risk of forcing the parties into a clandestine relationship, a potential consequence of which is the relationship's becoming public in the form of a future sexual harassment complaint or, at the very least, a grade appeal. A simple solution may be to ensure that neither party is in a position of power or control over the other, including reporting, supervision and evaluation structure. If the lab is sufficiently large, these duties can usually be transferred to another supervisory member. If small, the issue should be discussed and the student should be encouraged to consider withdrawing from the project if no acceptable arrangement can be made. Similar considerations should be made and conflict-of-interest policies should be consulted when recruiting a close relative of an existing lab member.

5.5 A FINAL WORD

A final consideration in developing undergraduate projects is the impact the experience will have on the student. While no effort should be made to disguise the realities of working in research, supervisors of undergraduate students should remain sensitive to the fact that students' first experiences in the research lab, whether good or bad, will undoubtedly have a long-term influence on their attitudes and perceptions toward research as a possible career path. A sour first experience could leave a lasting scar on many a potentially golden graduate student and future researcher. The PI and other supervisory personnel would be wise to keep this in mind while guiding undergraduate students through their formative research experience.

ACKNOWLEDGMENTS

We are grateful for the opportunity to work with students enrolled in the Independent Study Student and BSc Cooperative Education Programs at the University of Lethbridge. As well, RJW is thankful for support provided by the Natural Sciences and Engineering Research Council of Canada (Discovery Grant) and the Canada Research Chairs Program.

6 Sample Cards for Teaching Processing of Oilseeds and Cereals

Lawrence A. Johnson, Jeni Maiers and Darren Jarboe*

CONTENTS

6.1 INTRODUCTION

This chapter describes two sets of sample cards that are useful in teaching the processing and utilization of oilseeds and cereals. A companion website (www.ccur.info/samplecards) contains all the sample cards, book covers and posters discussed in this chapter that can be downloaded to your computer for printing. The sample cards provide flow diagrams of common methods used to process oilseeds and cereals into food and feed ingredients and products, biofuels and biobased products. Other sample cards show grading standards and types of oilseeds and cereals.

The authors first observed these kinds of teaching materials while taking a graduate cereal technology course taught by the late Professor Arlin Ward in the Department of Grain Science and Industry, Kansas State University. Jeni Maiers and Darren Jarboe have refined and expanded that concept with a focus on corn and soybean processing and products.

* Corresponding author.

6.1.1 Purpose

The materials described in this chapter have proven to be extremely useful in undergraduate and graduate courses, particularly for agronomy courses in oilseed and cereal crops; food science courses in fats and oils, cereal chemistry and technology, oilseed chemistry and technology, food carbohydrates, food proteins, and grain and food processing; and agricultural, biological and chemical engineering courses in engineering applications and grain and food processing. These materials have also been useful in training university outreach or extension personnel on ways to add value to crops and in introductory training of industry personnel, from operations to management, as well as in technical sales and product training applications.

A major benefit of using the sample cards is that students can see original materials, intermediate fractions and products, ingredients used in consumer products and final products that consumers readily recognize, all in the context of how they are produced. The sample cards show how fractions are recovered, converted into ingredients and used to manufacture finished products.

6.2 SAMPLE CARDS

There are 32 sample cards (www.ccur.info/samplecards/oilseeds.html) for oilseeds and 42 cards for cereals (www.ccur.info/samplecards/cereals.html). Additional cards will be added as need is identified. The sample cards are provided on the website in .pdf format for easy viewing and printing. The list of oilseed sample cards is shown in Table 6.1, while Table 6.2 lists the cards provided for cereals. Both lists are periodically updated on the website.

Each sample card provides process flow diagrams with spaces for respective samples to be mounted. The spaces are identified by boxes that are appropriately labeled for each sample. Figure 6.1 shows an example card for different oilseeds along with typical fatty acid compositions of oils recovered from these seeds. Figure 6.2 depicts soybean processing and shows how samples of dry products are mounted. Figures 6.3 and 6.4 show examples of soybean oil refining and degumming in which liquid samples have been incorporated. Alternatively, pictures of products rather than actual samples can be inserted in the .pdf file. Figure 6.5 shows an example of soymilk manufacturing with pictures of the samples inserted in the sample boxes. Pictures of the samples are not included on the website; however, detailed instructions on how to insert pictures are provided later in this chapter.

Samples of dry materials, such as grains, oilseeds flaked for extraction, ground meal and powdered ingredients, are placed in small polyethylene bags. Bags of three different sizes are used, depending on the sample size and the box size on the sample card. Liquid samples can also be displayed by placing the liquid into sample tubes, which in turn are placed in polyethylene bags and then stapled onto the cards. Suppliers, sizes and pictures of the bags and tube are shown in Table 6.3.

A long-reach stapler is recommended for mounting samples. We recommend using Swingline® stapler model #34121, which is ideal for stapling samples in the middle of a card or if there are multiple samples on a card.

TABLE 6.1
List of Sample Cards for Oilseeds

Oilseeds	
Sample Card Number	**Sample Card Title**
200	Oilseeds
201	Oilseeds
202	Types of Soybeans
203	Grading Standards for Soybeans
204	End-use Value Grading
205	Grain Cleaning
206	Soy Uses
207	Tofu
208	Soymilk
209	Soyfoods
210	Soybean Extraction
211	Soybean Extraction
212	Extraction Diagram
213	Oil Extraction Facility
214	Soybean Oil Refining
215	Degumming
216	Alkali Refining
217	Bleaching
218	Deodorization
219	Physical Refining
220	Meal Desolventization
221	Biodiesel
222	Soy Protein Products
223	Soy Protein Functional Properties
224	Soy Flour & Grits
225	Texturized Soy Proteins
226	Soy Protein Concentrates
227	Soy Protein Isolates
228	Soy Protein Texturization
229	Soy Protein Spinning
230	Soy Protein Fractionation
231	Extruders

Note: This list of sample cards for oilseeds can be found on the web at www.ccur.info/samplecards/oilseeds.html.

6.2.1 Workbooks and Covers

To create a workbook, completed sample cards can either be three-hole punched and placed into binders or compiled into a spiral-bound notebook. See Figure 6.6 for an example of each workbook. The binders work well for large amounts of sample

TABLE 6.2
List of Sample Cards for Cereals

Cereals	
Sample Card Number	**Sample Card Title**
100	Cereal Grains
101	Types of Wheat
102	Grading Standards for Wheat
103	Wheat in the United States
104	Wheat Flour Milling
105	Wheat Mill Flow Diagram
106	Pasta
107	The Oat Kernel
108	Oat Breakfast Cereals
109	Rice & Wheat Breakfast Cereals
110	Malting
111	Types of Corn
112	Types of Dent Corn
113	The Corn Kernel
114	Grading Standards for Corn
115	End-use Value Grading
116	Grain Storage
117	Grain Cleaning
118	Sweet Corn
119	Corn Drying & Storage
120	Mexican-Type Foods
121	Corn Hominy
122	Brewing
123	Fuel Ethanol Fermentation
124	Dry Corn Milling
125	Corn Dry Milling Flow Sheet 1
126	Corn Dry Milling Flow Sheet 2
127	Corn Wet Milling
128	Wet Corn Milling
129	Corn Refining
130	Corn Oil Processing & Refining
131	Different Starches
132	Amylose & Amylopectin
133	Starch Granules
134	Modified Corn Starches
135	Modified & Derivatized Starches
136	Chemical Conversion to Glucose
137	Enzymatic Conversion to Dextrose
138	Enzymatic Conversion to Fructose
139	Different D.E. Corn Syrups
140	Corn Zein Isolation
141	Corn Sweetners

Note: This list of sample cards for cereals can be found on the web at www.ccur.info/samplecards/cereals.html

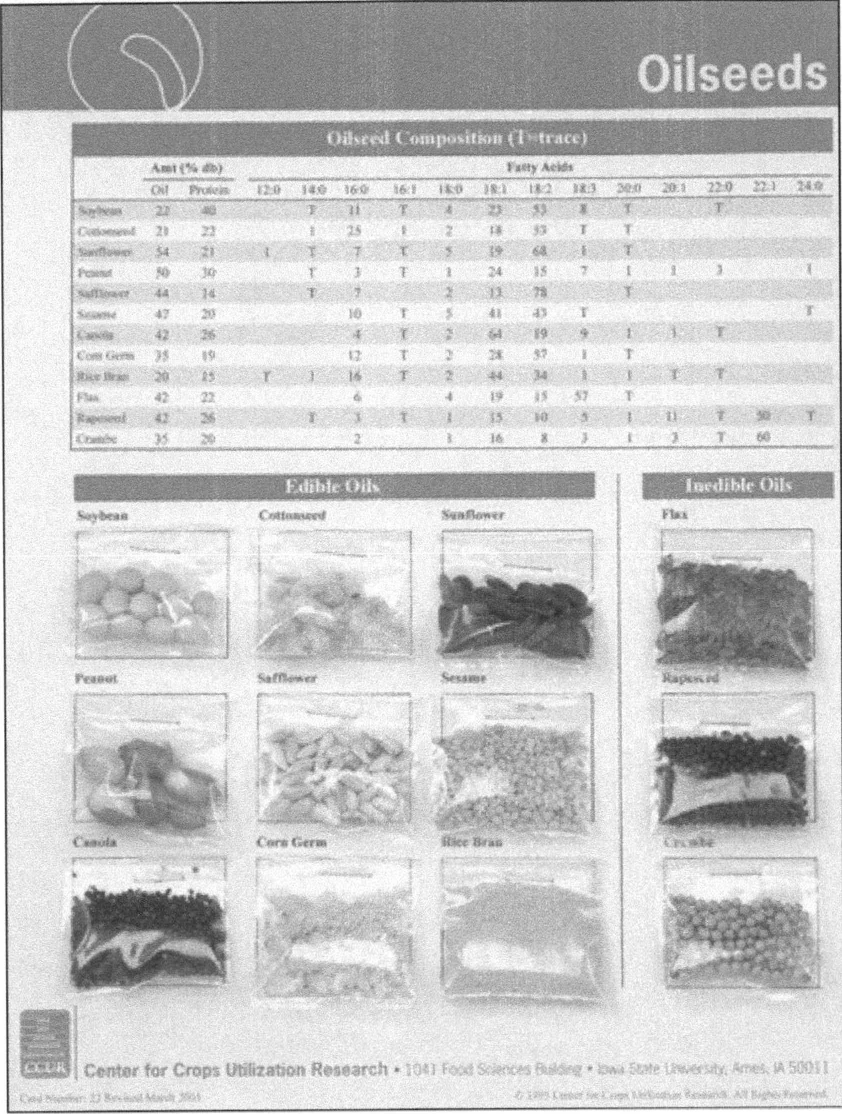

	Amt (% db)		Fatty Acids												
	Oil	Protein	12:0	14:0	16:0	16:1	18:0	18:1	18:2	18:3	20:0	20:1	22:0	22:1	24:0
Soybean	22	40		T	11	T	4	23	53	8	T		T		
Cottonseed	21	22		1	25	1	2	18	53	T	T				
Sunflower	54	21	1	T	7	T	5	19	68	1	T				
Peanut	50	30		T	3	T	1	24	15	7	1	1	3		1
Safflower	44	14		T	7		2	13	78		T				
Sesame	47	20			10	T	5	41	43	T					T
Canola	42	26			4	T	2	64	19	9	1	1	T		
Corn Germ	35	19			12	T	2	28	57	1	T				
Rice Bran	20	15	T	1	16	T	2	44	34	1	1	T	T		
Flax	42	22			6		4	19	15	57	T				
Rapeseed	42	26		1	3	T	1	15	10	5	1	11	T	50	T
Crambe	35	20			2		1	16	8	3	1	3	T	60	

FIGURE 6.1 Oilseeds sample card with mounted samples.

cards (10 or more). We recommend using binders with 3- to 5-inch capacities holding 8.5- by 11-inch paper. Spiral-bound notebooks are ideal for 10 cards or less. The spiral combs come in a variety of sizes, typically ranging from a quarter of an inch to 2 inches. Most copy centers will be able to bind the spiral-bound notebooks, and they will be able to help pick out the correct comb size.

Included on the website are covers for binders and spiral-bound notebooks. Binders are available that have a plastic sleeve covering, allowing the cover to be placed inside. A cover for the binder spine is also included.

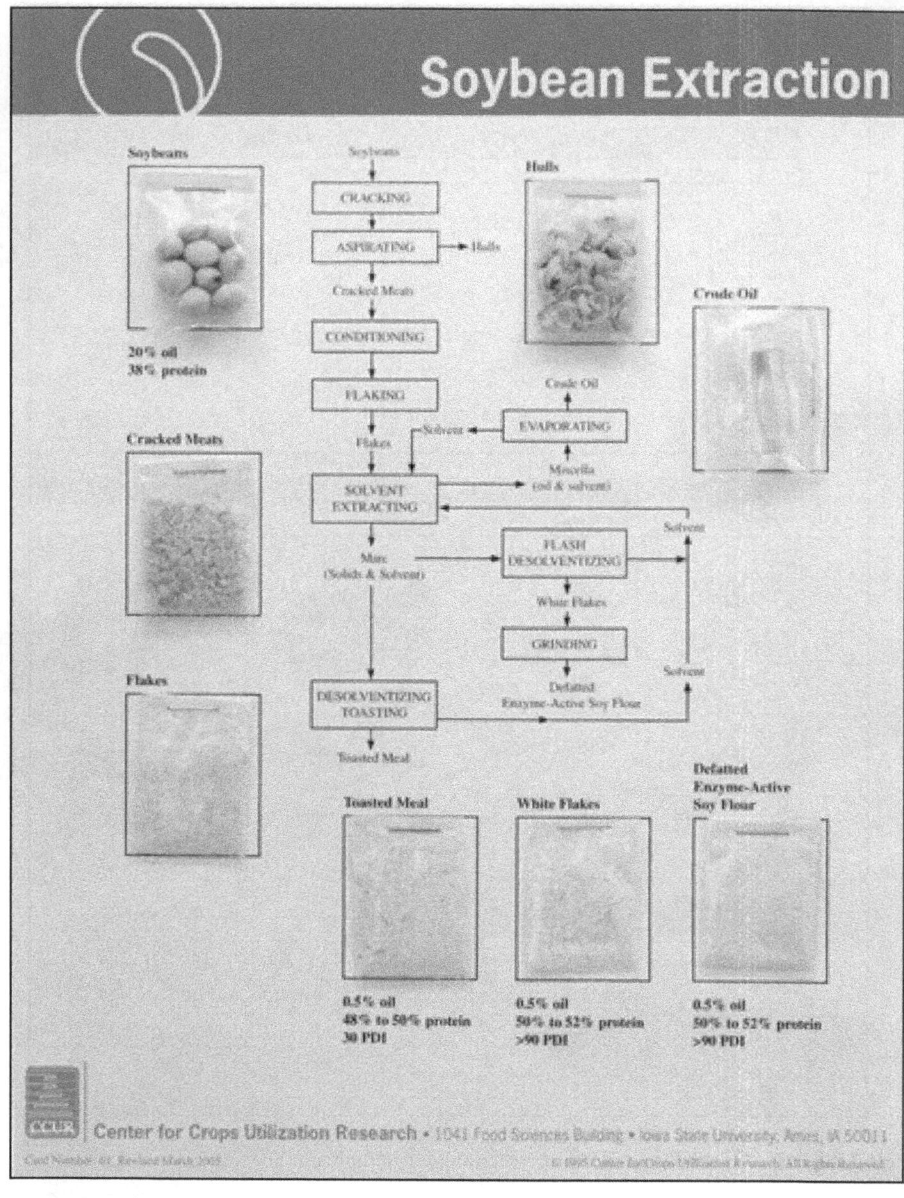

FIGURE 6.2 Soybean extraction sample card.

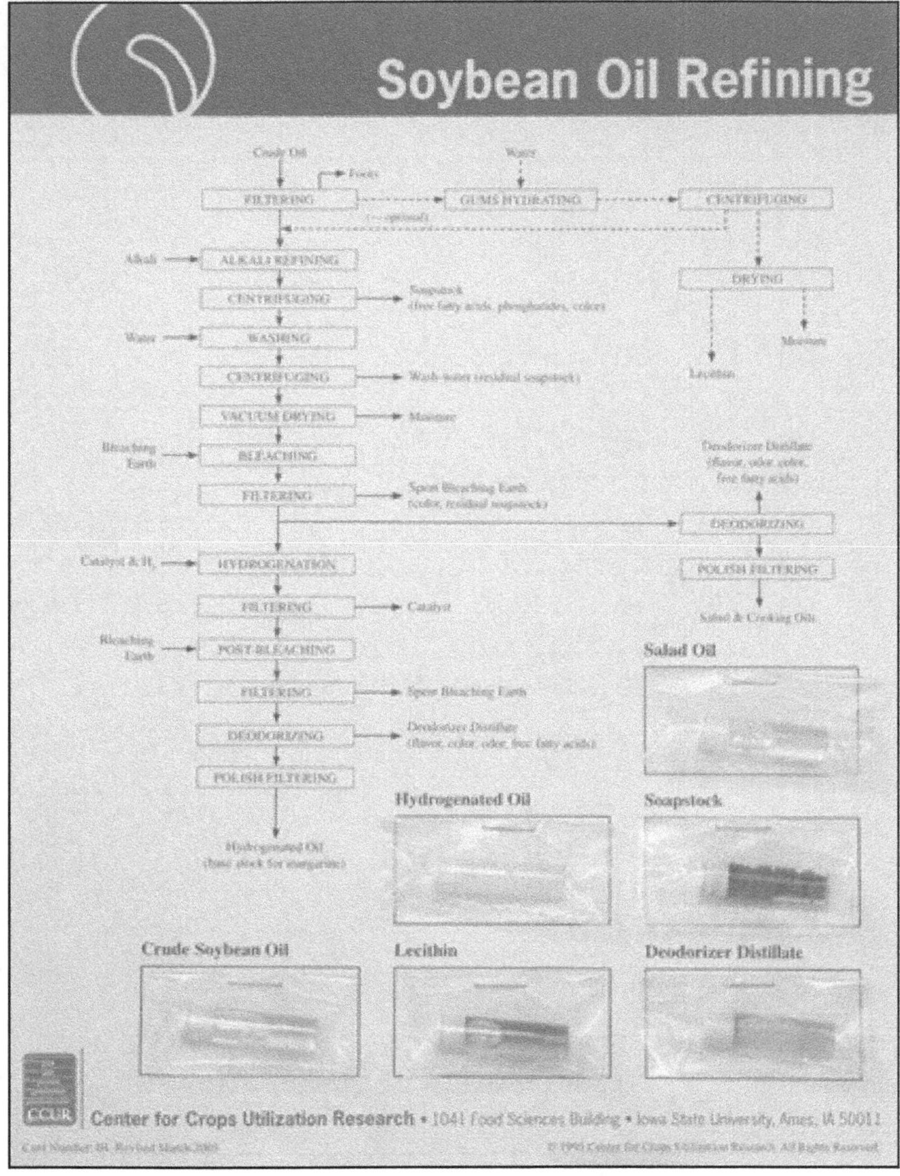

FIGURE 6.3 Soybean oil refining sample card with liquid samples.

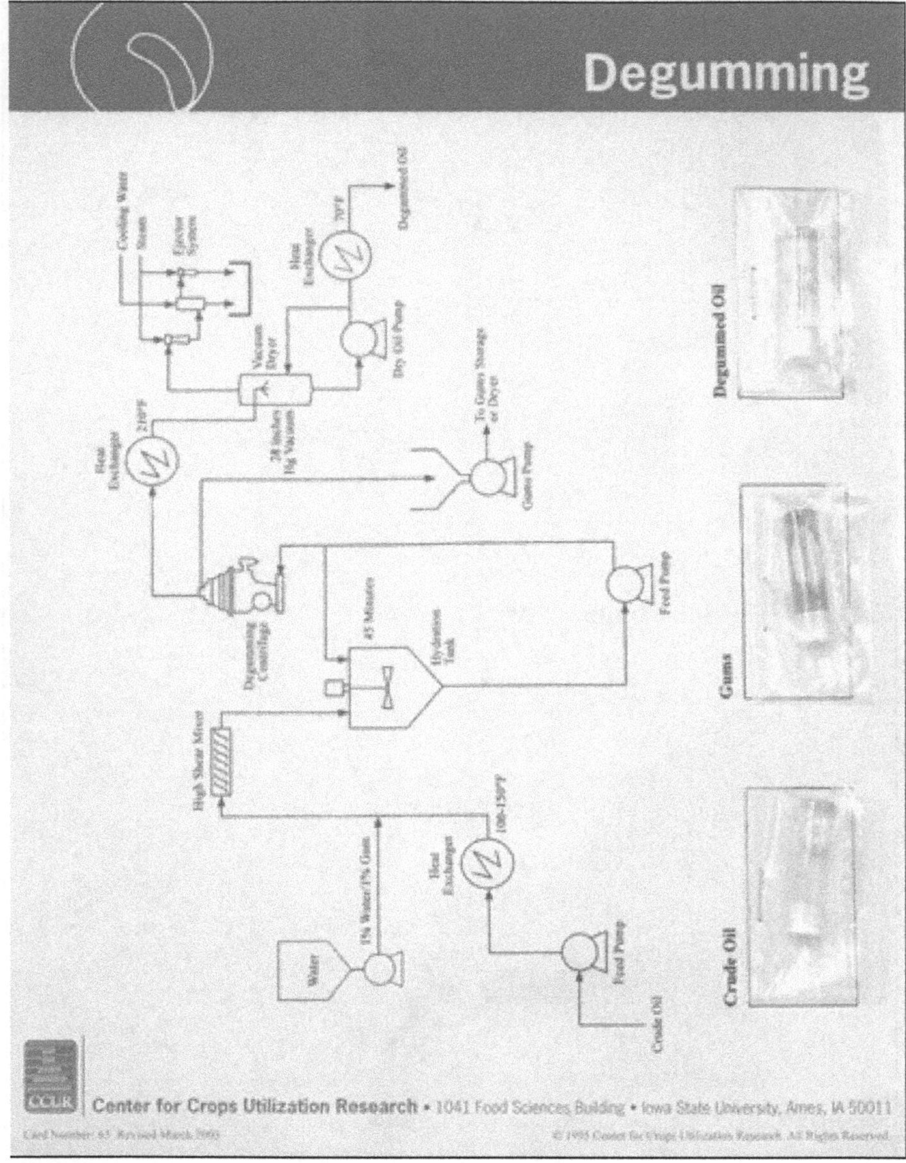

FIGURE 6.4 Degumming sample card with liquid samples. (Degumming process redrawn from diagram provided by Delaval Separator Co., Sulllivan Systems, Inc., Larkspur, California.

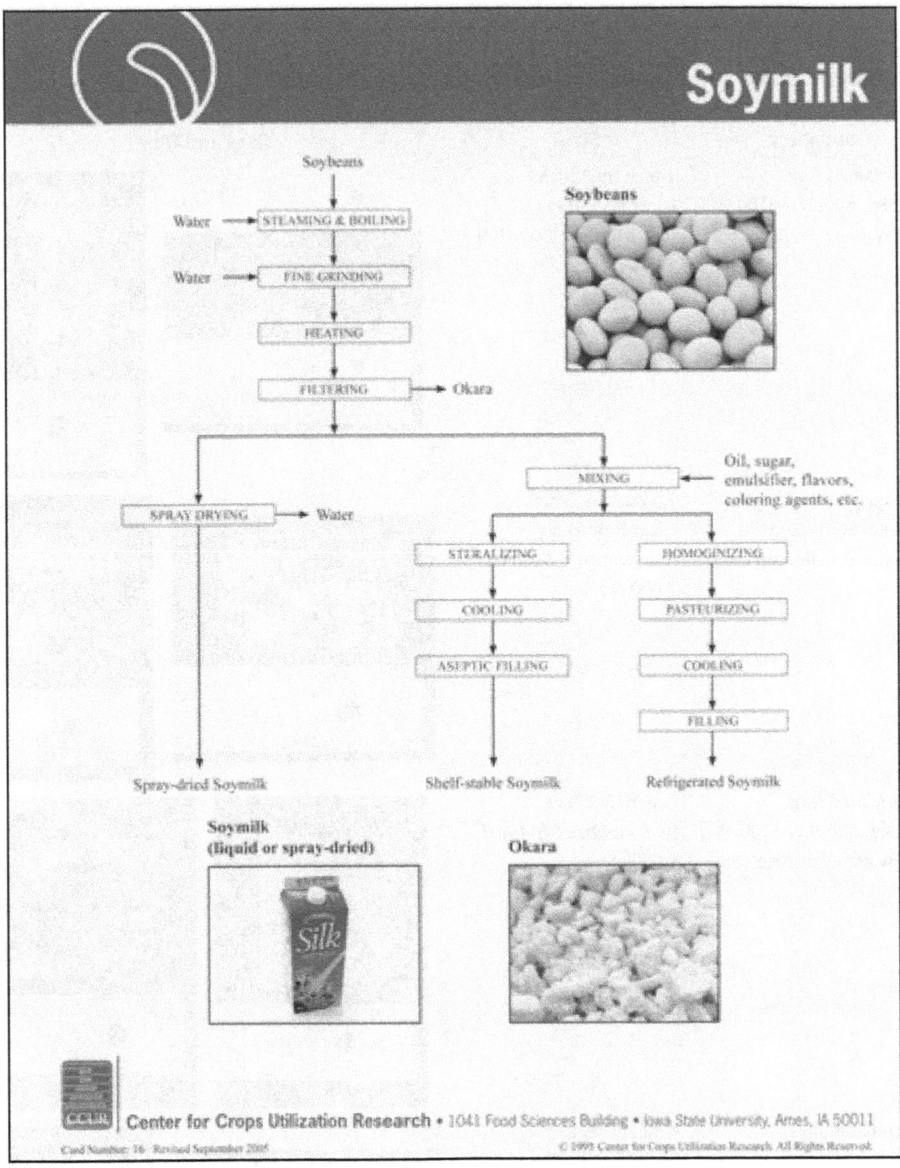

FIGURE 6.5 Soymilk sample card with pictures inserted as samples.

TABLE 6.3
Suppliers, Sizes and Photos of the Sample Bags and Tube

Suppliers	Sizes	Bags and Tube
Associated Bag Phone: 800-926-6100 www.associatedbag.com	1 inch by 2 inch Clear Poly Bags Item number: 28-4-400 1,000 pieces	 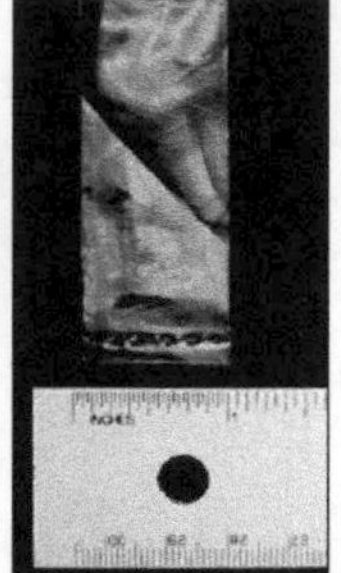
Associated Bag Phone: 800-926-6100 www.associatedbag.com	1.5 inch by 2 inch Clear Poly Bags Item number: 28-4-402 1,000 pieces	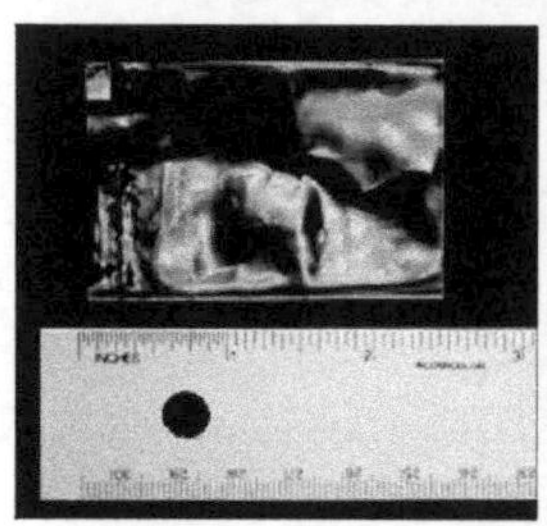 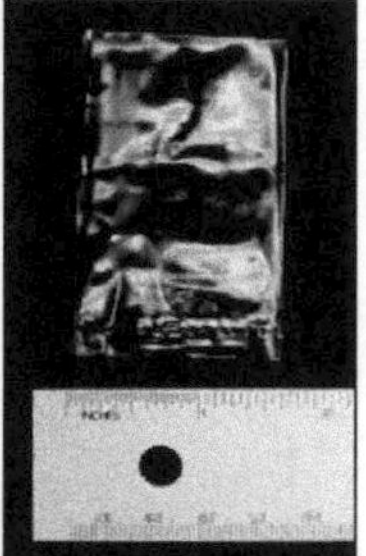
Associated Bag Phone: 800-926-6100 www.associatedbag.com	2 inch by 2 inch Item number: 28-4-405 1,000 pieces	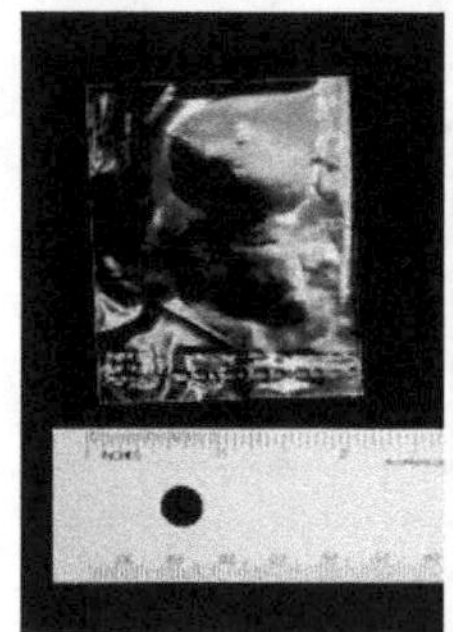 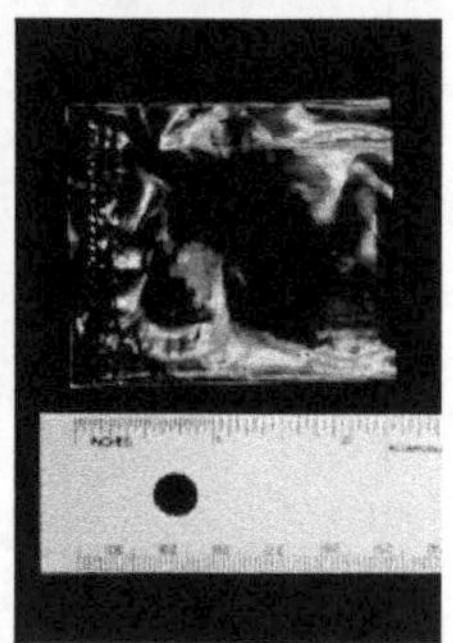
Chromatography Research Supplies, Inc. Telephone: 800-327-3800 (U.S. only) 502-491-6300 Fax: 502-491-3390 www.chromres.com	8 mm capacity Shell Vial & Plug Item number: 083042 200 per pack	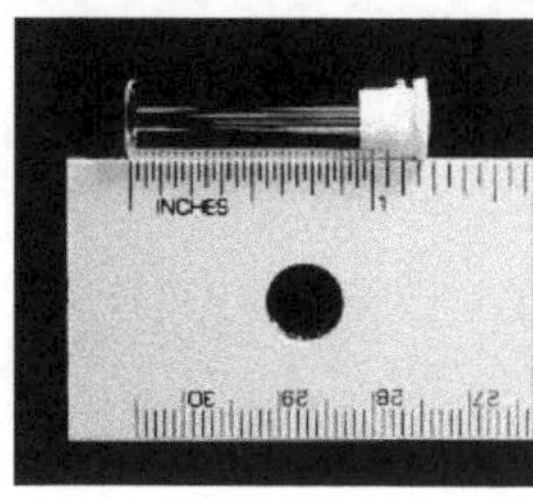 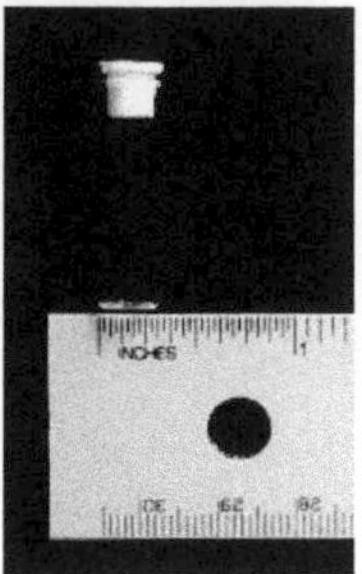

FIGURE 6.6 Three-hole binder (left) and spiral-bound notebook.

6.2.2 Inserting Pictures

Adobe® Acrobat® 7.0 Professional or later is required to insert pictures into a .pdf file. Pictures of the samples are not included on the website. Listed below are steps for inserting pictures into a .pdf document.

6.2.2.1 Copying the Picture

1. Gather the respective pictures on the computer.
   (a) Open the picture in a photo editing program such as Adobe Photoshop.

   or

   (b) Insert the picture into another program such as Microsoft® Word (Figure 6.7).
2. Select the picture and choose Edit>Copy.

6.2.2.2 Inserting the Picture into a .pdf Document

1. Open the sample card.
2. Click the View menu and find the option called Snap to Grid. If there is a checkmark, deselect it (Figure 6.8). This will allow greater freedom to move the picture around in the document.
3. Go to the toolbar and select Commenting>Stamp Tool>Paste Clipboard Image as Stamp Tool (Figure 6.9).

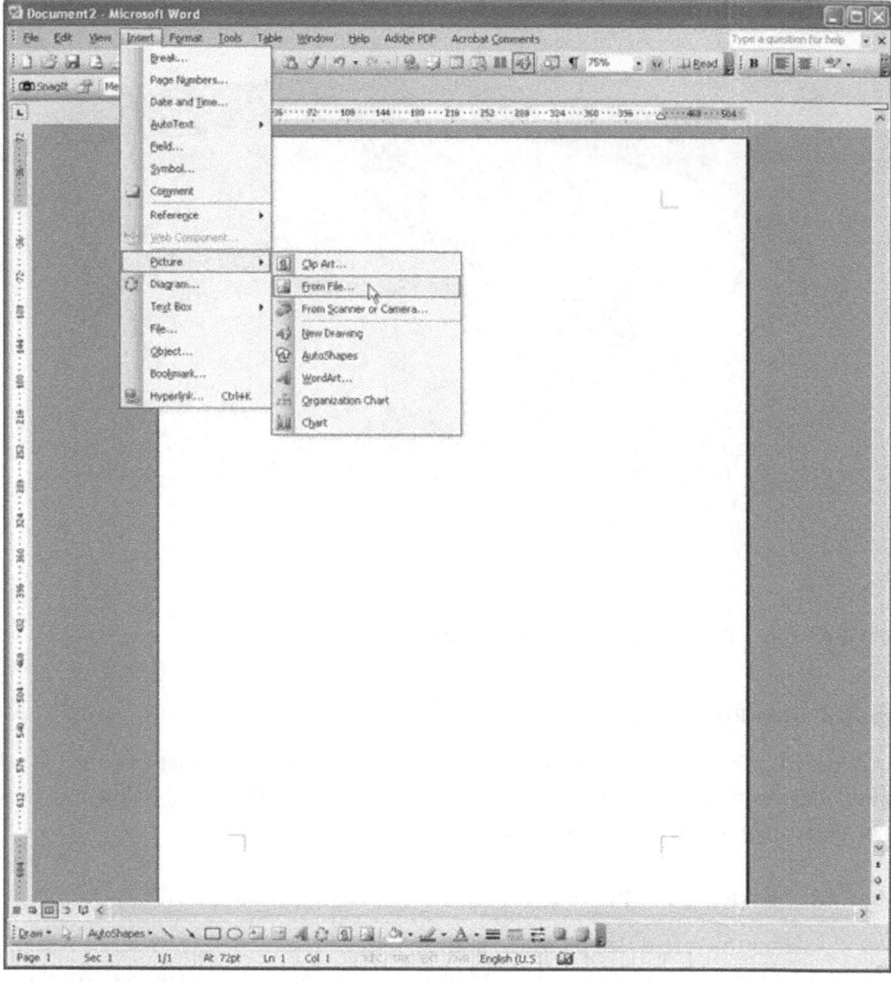

FIGURE 6.7 Screen shot showing how to insert a picture in Microsoft Word.

4. Next, click anywhere in the document. The picture is now inserted in the document.
5. To move the picture, select the Hand Tool. Click and hold the mouse to move the picture around (Figure 6.10).
6. To resize the picture, select the Hand Tool, and while holding down the Shift key, drag one of the picture's handles to resize it (Figure 6.11).
7. To delete a picture, right click (Windows) or Control-click (Mac OS) the picture and select Delete from the menu (Figure 6.12).

6.2.3 Paper

A few options exist when selecting paper for the sample cards. Regular 8.5- by 11-inch white copy paper will work if the cards contain only pictures of the samples.

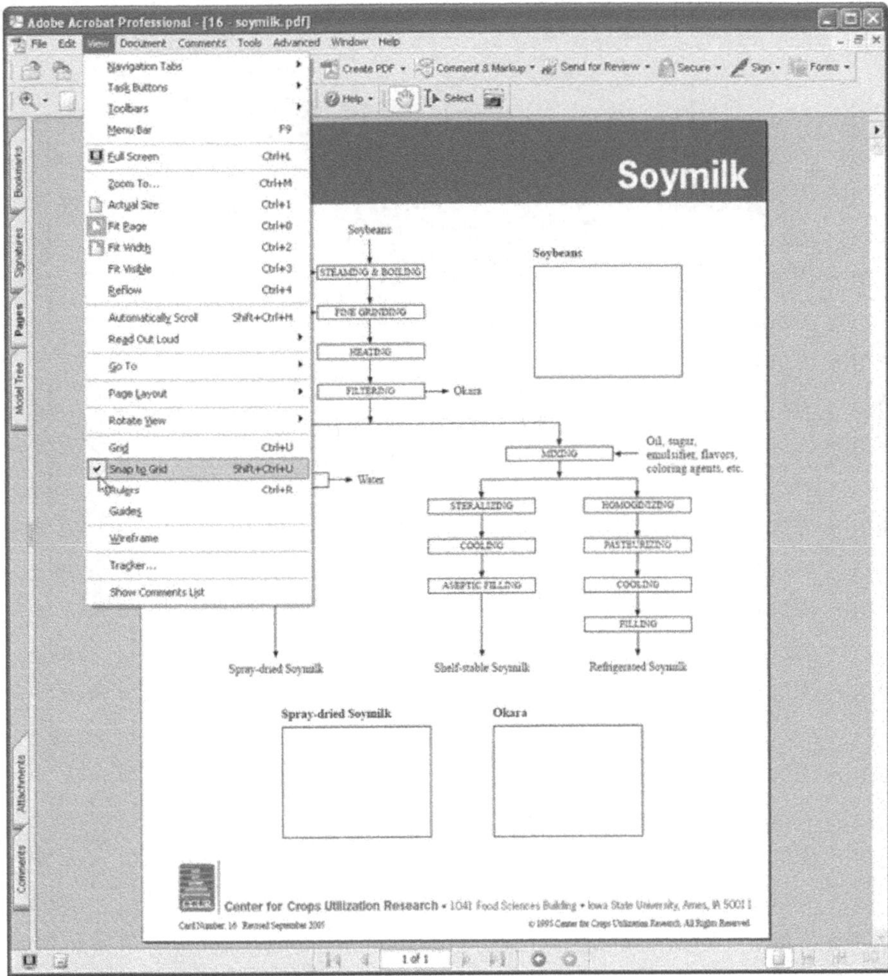

FIGURE 6.8 Screen shot showing how to deselect Snap to Grid.

However, using 70-pound white or off-white paper will achieve better, more professional results.

Sample cards with mounted samples require heavier paper to withstand the weight of the samples. Either cover or card stock will work for this application.

6.2.4 Printing

The sample cards are designed for color and borderless printing (printing that reaches the edges of the paper, see Figure 6.5 for an example). However, the .pdf files can still be printed on any desktop printer, either in color or black and white. Most desktop printers will not be capable of borderless printing, resulting in a white border around the card. Also, make sure to check the printer's manual to see if it can print on cover or card stock. Most desktop printers cannot use these heavy papers.

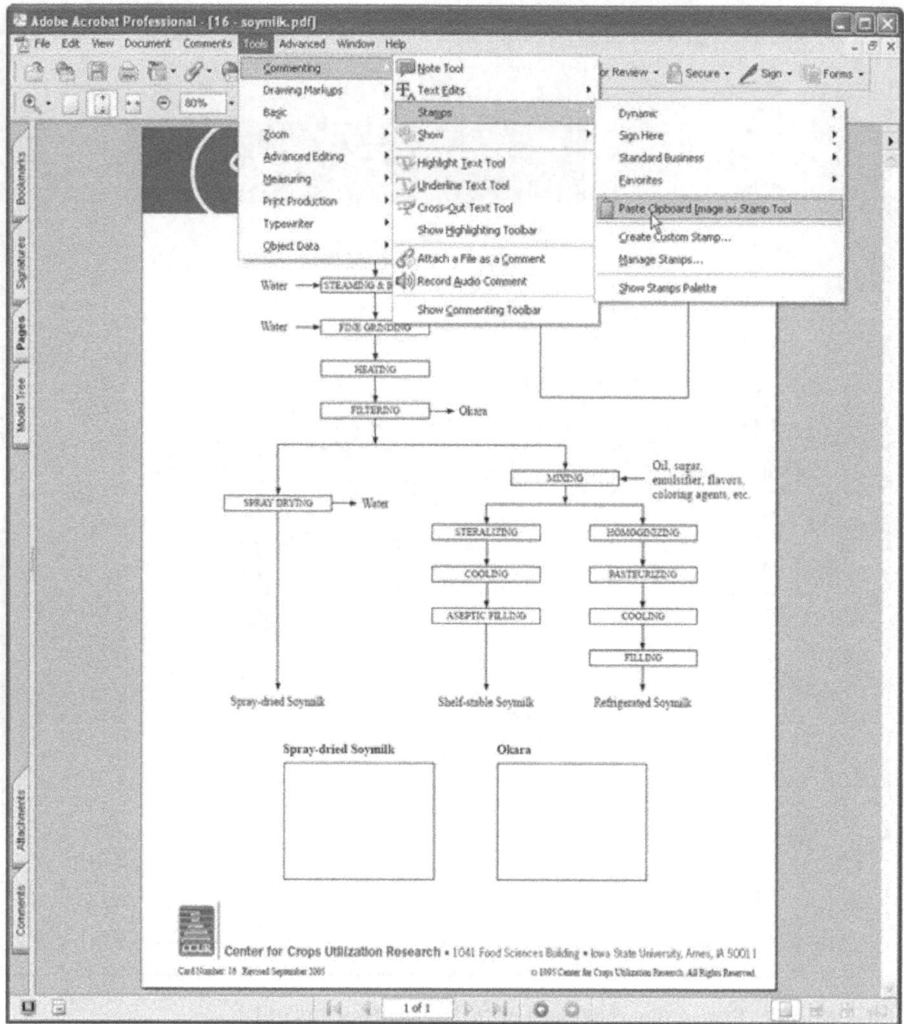

FIGURE 6.9 Screen shot showing where to locate the Stamp Tool.

If borderless printing is preferred, you have two options. The first is to take the .pdf files to a commercial printer. Make sure to request borderless printing and cover or card stock. The second option is to purchase a desktop printer capable of borderless printing on heavier paper. The Hewlett-Packard® DeskJet 9800 is an example of a printer that can achieve borderless prints on cover or card stock.

To print the workbook covers, follow the same principles used for the sample cards. One exception, however, is the binder cover. The binder cover size is 11.25 inches wide by 11 inches high, which is beyond the capabilities of most standard desktop printers. Instead, use the spiral-bound cover, which will print on all desktop printers. These covers are also ready for printing at a commercial printer.

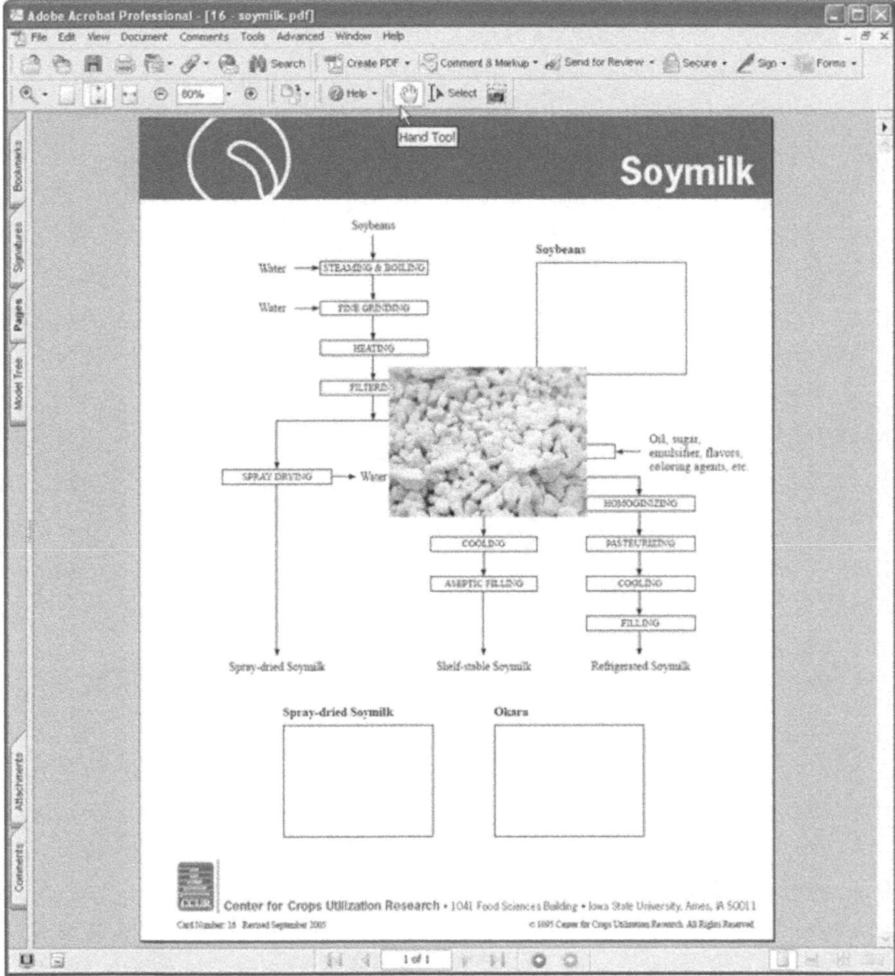

FIGURE 6.10 Screen shot showing the Hand Tool.

6.3 SUPPLEMENTARY MATERIALS

We often supplement the sample cards with additional materials such as references on oilseeds processing and utilization by Erickson,[1] Wan and Farr[2] and Johnson[3] and references by White and Johnson,[4] Kulp and Ponte[5] and Wrigley, Corke, and Walker[6] on cereals processing and utilization. These references provide much more detail than can be included on the sample cards. The *Soya & Oilseed Bluebook*,[7] published annually by Soyatech, Inc., provides oilseed statistics on production, processing, products and utilization. The tables and graphs can be downloaded from www.soyatech.com after subscribing to the publication. The U.S. Department of Agriculture Economic Research Service website (www.ers.usda.gov) can provide additional economic information.

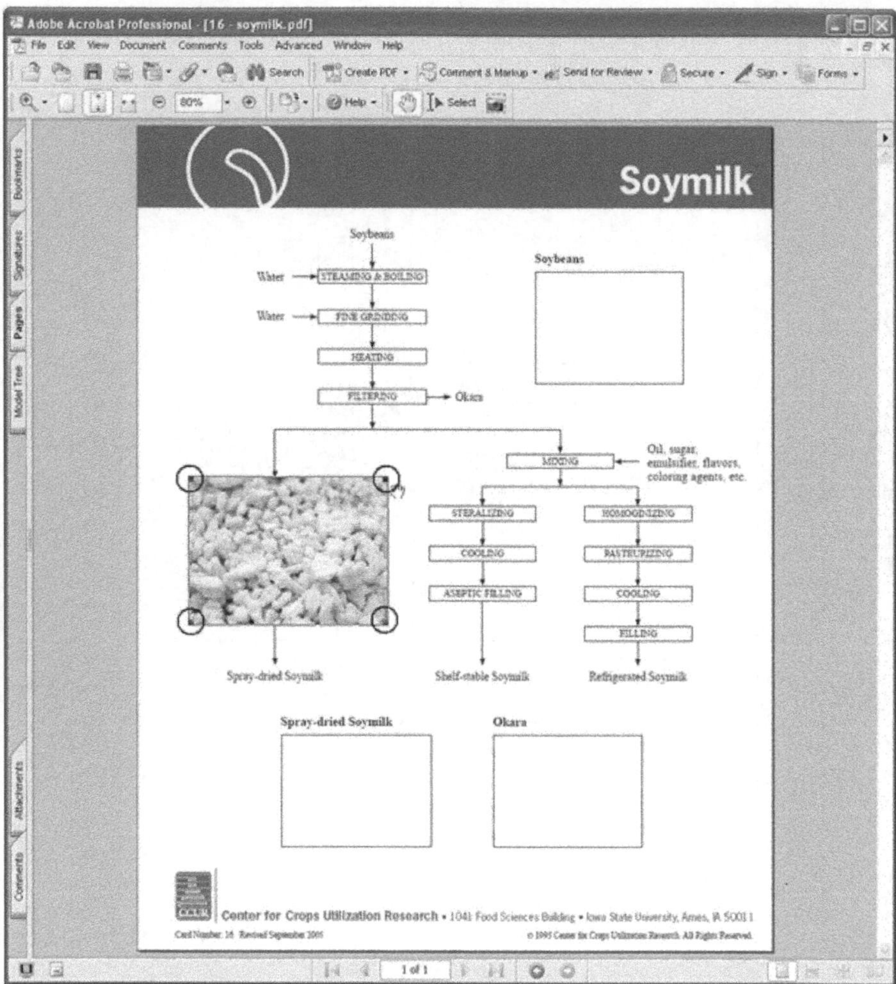

FIGURE 6.11 Screen shot showing the resize handles.

6.4 POSTERS

The .pdf files of two posters are also provided on the website: one on soybean processing and utilization (Figure 6.13) and another poster on corn processing and utilization (Figure 6.14). The posters depict how these crops are fractionated and converted into food ingredients, biofuels and bio-based products. We have used the posters as placemats for dinner parties, teaching materials in the sample cards workbook and laminated posters hung in high-traffic areas. The posters are 11 inches high by 17 inches wide. They will easily reduce for printing on a standard desktop printer (8.5 by 11 inches up to 11 by 14 inches). We do not recommend enlarging the posters beyond 11 by 17 inches.

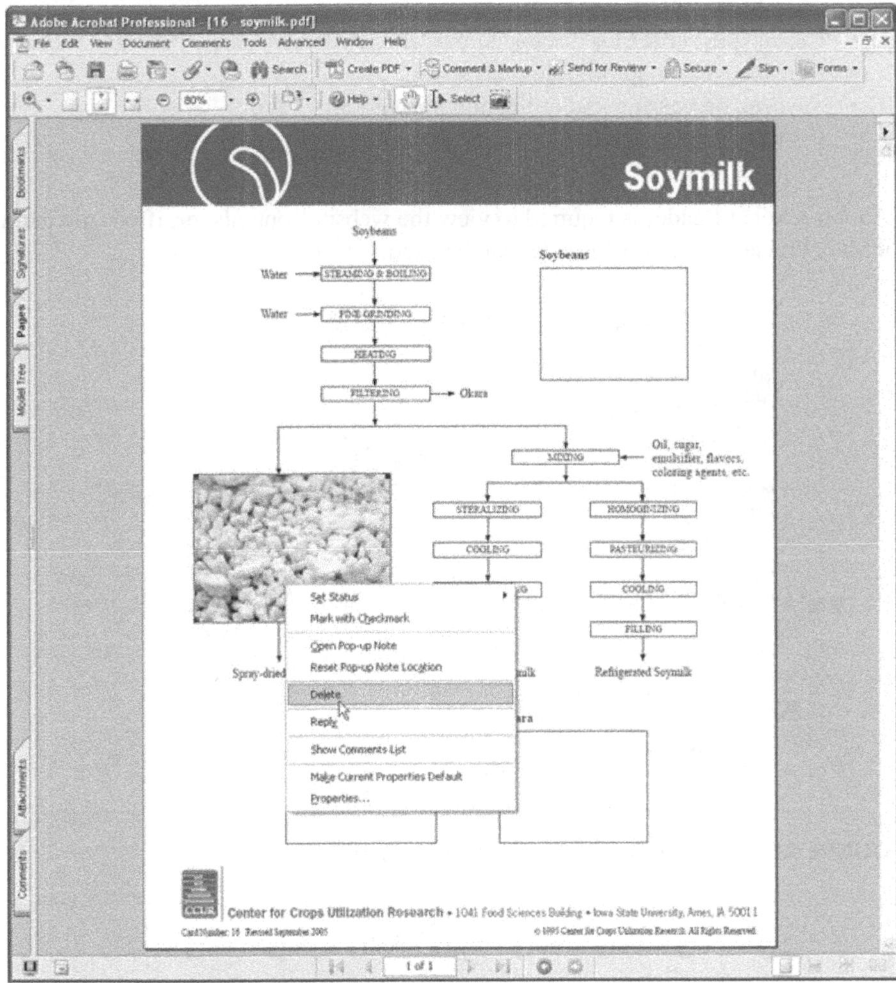

FIGURE 6.12 Screen shot showing how to delete a picture.

6.5 WEBSITE CONTENTS AND COMPUTER PROGRAM

Included on the website are the following:

- List of oilseed sample cards
- List of cereal sample cards
- 32 oilseed sample cards
- 42 cereal sample cards
- Oilseed cover for three-ring binder
- Oilseed spine cover for three-ring binder
- Oilseed cover for spiral-bound notebook

- Cereal cover for three-ring binder
- Cereal spine cover for three-ring binder
- Cereal cover for spiral-bound notebook
- Soybean processing and utilization poster
- Corn processing and utilization poster

Adobe Acrobat Reader is required to view the website contents, or, if you are using a Mac, Preview or Acrobat Reader can be used.

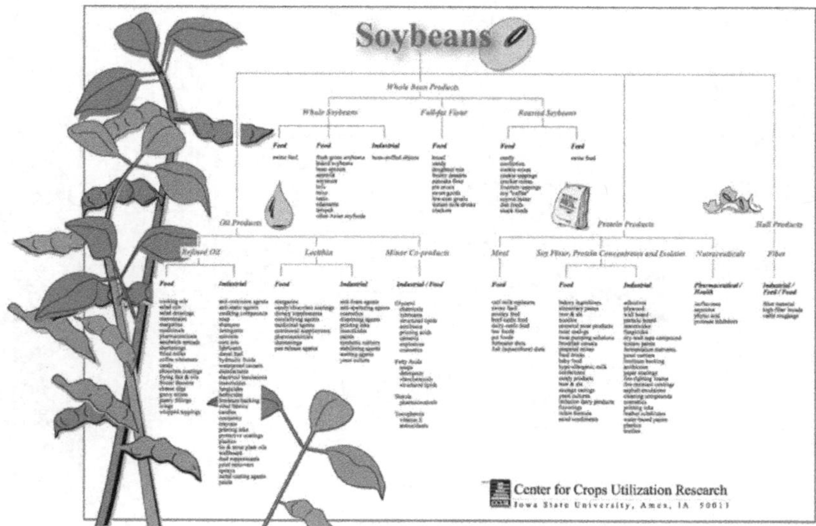

FIGURE 6.13 Soybean processing and utilization poster.

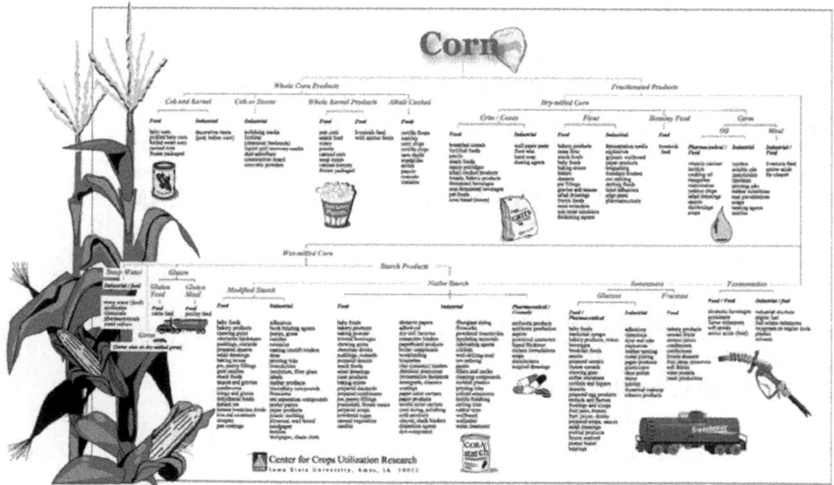

FIGURE 6.14 Corn processing and utilization poster.

REFERENCES

1. Erickson, D. (ed.). 1995. *Handbook of Soybean Processing and Utilization*, AOCS Press, American Oil Chemists Society, Champaign, IL.
2. Wan, P.J. and W. Farr (eds.). 2000. Recovery of Fats and Oils from Plant and Animal Sources. In *Introduction to Fats and Oils*, AOCS Press, American Oil Chemists Society, Champaign, IL.
3. Johnson, L.A. 2002. Recovery, Refining, Converting, and Stabilizing Edible Fats and Oils. In *Food Lipids* (second edition), edited by C. Akoh and D. Min. Marcel Dekker, New York. pp. 223–273.
4. White, P.J. and L.A. Johnson (eds.). 2003. *Corn Chemistry and Technology*, AACC Monograph Series. American Association of Cereal Chemists, St. Paul, MN.
5. Kulp, K. and J.G. Ponte, Jr. (eds.). 1999. *Handbook of Cereal Science and Technology* (second edition), Marcel Dekker, New York. pp. 31–80.
6. Wrigley, C., H. Corke and C. Walker (eds.). 2004. *Encyclopedia of Grain Science.* Academic Press, Oxford, UK.
7. *Soya & Oilseeds Bluebook.* Soyatech, Inc., Bar Harbor, ME.

7 Biotechnology and Crop Improvement in Agriculture

David Hildebrand

CONTENTS

7.1 INTRODUCTION

In the general sense, biotechnology is the application of technology to biological systems such as crop plants. Plant breeding can be considered a biotechnology, but when most people talk of crop biotechnology they are referring to crops genetically engineered using modern recombinant DNA techniques.

7.1.1 History

The history of crop breeding is intertwined with the history of civilization. Before crops were domesticated and farming began, people needed to spend most of their time hunting and foraging for their food and could not spend time on art, architecture and engineering, nor could they live in cities with high populations. Plant and animal domestication and agriculture changed all that, allowing a surplus of food to be produced to feed people who spent little or no time gathering or producing food. Crop domestication and formation of civilizations began ~10,000 years ago in southwest Asia with wheat, in east Asia with rice and in meso-America with corn—the plants that remain the three most important crops to this day, although domestication and breeding of dogs is thought to predate crop plant domestication and breeding (Table 7.1) (Doebley, 2006; Fernandez et al., 2006; Hanotte et al., 2002; Konishi et al., 2006; Larson et al., 2005; Li et al., 2006; Savolainen et al., 2002; Weiss et al., 2006).

Crop improvement and the art of plant breeding goes back to the beginning of crop domestication. The science of genetics and breeding began with Mendel's classical experiments some 150 years ago. Mendel reported his inheritance experiments in 1866, and the foundation of molecular genetics started just 3 years later in 1869 when Miescher discovered a weak acid abundant in the nuclei of white blood cells that came to be known as DNA (Hartl and Jones, 2005). General features of chromosomes became known by around 1900, and, by the 1920s, several lines of evidence indicated a relationship between chromosomes and DNA (Table 7.1).

Genetic transformation was first reported in *Streptococcus* in 1928, and in 1944 it became understood that DNA was responsible for the genetic transformation. Experiments published by Hershey and Chase in 1952 substantiated that DNA is passed from one generation to the next in a virus. In 1953, Watson and Crick proposed that DNA is structured as a twisted double helix, which was our first understanding of the three-dimensional structure of DNA. It was elucidated that the genetic code consists of four letters of the four different bases of DNA: adenine (A), cytosine (C), guanine (G) and thymine (T). In double-stranded DNA, A pairs with T and C pairs with G via hydrogen bonding.

The "Central Dogma" of gene expression is that DNA is transcribed into RNA and RNA is translated into protein. "Words" of the genetic code are three letters known as a codons and each codon encodes a specific amino acid. For example, the codon ATG is transcribed into the corresponding codon in RNA, AUG, and translated into the amino acid methionine. The field of recombinant DNA goes back

TABLE 7.1
Timeline of Biotech and Crop Improvement

15,000 years ago (bp)	Domestication of dogs
8–15,000 bp	Domestication of cattle, goats, pigs and sheep
10,000 bp	First crops domesticated
10,000 bp until 1800s	Most modern crops created through the art of plant breeding
1800s	Beginnings of the science of genetics and its application to crop improvement
1866	Mendel describes genetic inheritance
1900	General features of chromosomes become known
1928	Genetic transformation reported
1944	DNA found to be responsible for genetic transformation
1940s and 50s	Crop hybrids such as corn developed; seed industry develops
1950s	Watson and Crick describe the structure of DNA
1970s	Recombinant DNA techniques developed
1980s	Genetic engineering (GE) of higher plants and animals developed
1990s	GE crops widely grown by farmers
1990s+	GE of oilseed crops developed for different end uses of the oil

to the discovery of restriction enzymes in the 1970s, allowing the cutting and splicing of specific DNA fragments. Werner Arber was awarded a Nobel prize in 1978 for elucidation of restriction enzymes and their uses. The use of recombinant DNA was rapidly adopted for microbial genetics and breeding studies in the 1970s. It became known that microorganisms such as bacteria and yeast can take up and express DNA from virtually any source including higher animals and plants. An early commercial application of recombinant DNA techniques was the production of human insulin in microorganisms, and diabetics have been using recombinant human insulin produced in microbes ever since (http://www.gene.com/gene/about/corporate/history/timeline/index.jsp).

It was not until the 1980s that recombinant DNA techniques were applied to higher eukaryotic organisms such as higher plants and animals. Palmiter et al. (1983) reported on the development of transgenic mice. Studies of the tumor-inducing soil bacteria *Agrobacterium tumefaciens* led to the discovery that this organism naturally transfers genes into the genomes of plants (Zambryski et al., 1980). The DNA that is transferred, termed T-DNA encoded on tumor-inducing plasmids (Ti plasmids), was found to encode genes that induce the tumors in the transformed plant tissues and it was found that these genes can be replaced with other genes of interest, thus disarming the Ti plasmid and allowing the use of this natural plant transformation vehicle for plant genetic engineering (Figure 7.1). The first reproducible method for plant transformation and regeneration was reported by Rob Horsch in 1985 (Horsch et al., 1985).

This led to the development of so-called crop "genetic engineering," and the first "genetically engineered, or GE" crop introduced into the marketplace was the Flavr Savr tomato by Calgene Inc. in 1994 (http://dragon.zoo.utoronto.ca/~jlm2000/T0501D/introduction.html; http://www.geo-pie.cornell.edu/crops/tomato.html). The first wide-scale production GE crops began in 1996 with the introduction of Bt and Roundup Ready™ crops, especially corn, soybeans, cotton and canola. Such GE crops have been enormously popular with farmers, and in 2006 in the U.S., 89% of the soybean crop representing ~27 million ha, 83% of the cotton crop representing ~5.3 million ha and 61% of the corn crop representing ~19 million ha were GE or biotech crops (http://usda.mannlib.cornell.edu/reports/nassr/field/pcp-bba/acrg0606.pdf).

7.2 MAJOR BIOTECH CROPS IN PRODUCTION BY FARMERS

7.2.1 Bt Crops

Bt crops contain a delta endotoxin protein that harms only members of a specific insect family such as Lepidoptera larvae, in particular, European corn borer. Growers use Bt corn as an alternative to spraying insecticides for control of European and southwestern corn borer. Bt crops can save farmers pesticide costs and increase yields. They are compatible with biological insect control and are considered more environmentally friendly than use of broad-spectrum insecticides, which can be reduced or eliminated with the use of Bt crops (http://www.uky.edu/Ag/Entomology/entfacts/fldcrops/ef130.htm).

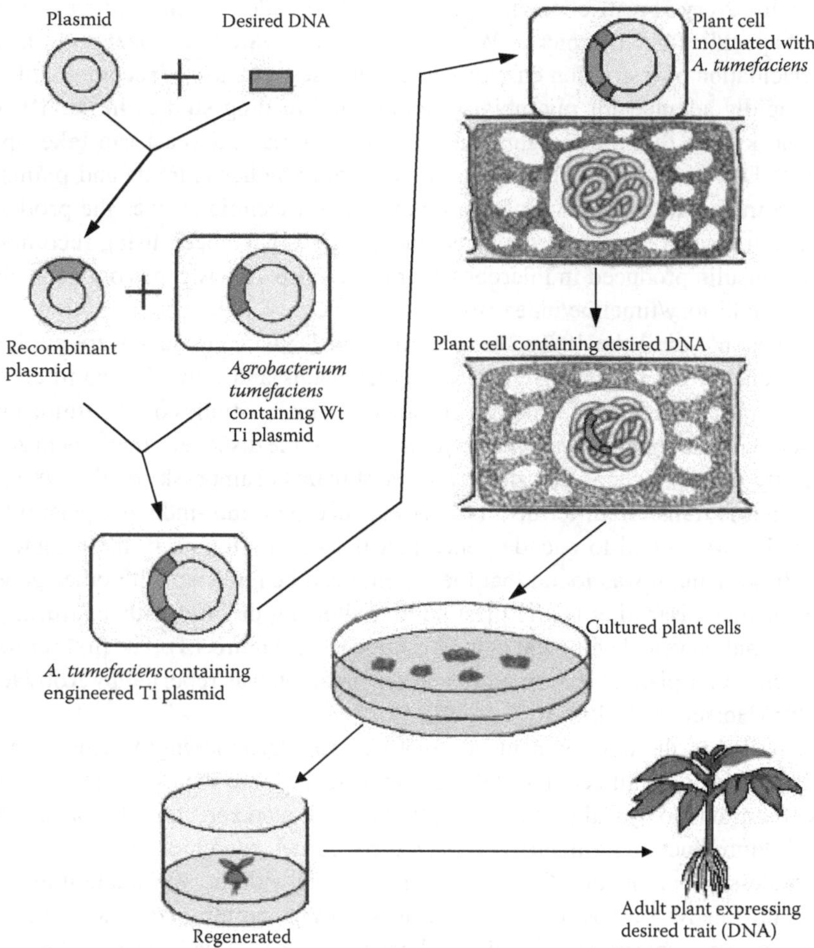

FIGURE 7.1 General process for plant genetic engineering (inserting foreign genes into plant cells). A plasmid containing DNA is cut with a restriction enzyme and DNA of the desired gene (gray) is inserted. The desired gene is then inserted into Ti (tumor-inducing) plasmid naturally found in *A. tumefaciens.* Plant cells are inoculated with *A. tumefaciens* containing the engineered Ti plasmid + the desired DNA. *A. tumefaciens* transfers the desired DNA into plant chromosomes. Plantlets with the desired trait are then regenerated.

The Bt gene in Bt crops was isolated from the naturally occurring soil bacterium *Bacillus thuringiensis.* Bt insecticides have been used since the 1960s and are considered among the safest and most environmentally benign pest control agents because of their very high selectivity, excellent safety record and ready biodegradability. Bt does not harm insects of nontarget families or other organisms (http://npic.orst.edu/factsheets/BTgen.pdf). It is desirable, however, only to target pest species within target families and minimize effects on other species.

The Lepidoptera family includes not only corn borers but also butterflies. A preliminary small-scale lab study published as a note in *Nature* in June 1999 suggested that there might be impacts of Bt corn pollen on monarch butterfly caterpillars, which feed on milkweed leaves. In this experiment an impact on young monarch butterfly caterpillars was observed when they were force fed milkweed leaves with a thick coating of pollen from a specific rare type of Bt corn with unusually high Bt protein levels in the pollen. This prompted a series of field studies to determine if there is a detrimental impact of Bt corn on monarch butterflies under natural environmental conditions. The results of five of these studies conducted around the U.S. and Canada are published in the *Proceedings of the National Academy of Sciences of the United States of America* and are summarized by the USDA at http://www.ars.usda.gov/is/br/btcorn/index.html#bt9. It was concluded that Bt corn causes no immediate harm to monarch butterflies under natural field conditions. It is worth noting that although Bt crops are not as precise in their targeting of serious crop pests as is ideally desired, they are far more precise than alternative pest control methods developed so far.

7.2.2 Roundup Ready Crops

Roundup™ is the most successful herbicide in history. A key step in the synthesis of the aromatic amino acids, phenylalanine (Phe), tyrosine (Tyr) and tryptophan (Trp), is the condensation of phosphoenol pyruvate (PEP) with shikimate 3-phosphate, forming 5-enoylpyruvylshikimate 3-phosphate (EPSP) catalyzed by EPSP synthase (Figure 7.2). The active ingredient of Roundup is glyphosate, which is structurally similar to PEP (Figure 7.3). Glyphosate inhibits EPSP synthase, preventing the formation of the essential amino acids Phe, Trp and Tyr killing the plant tissues. All living organisms use 20 different amino acids as building blocks to make their proteins. Animals, including humans, are able to synthesize only 10 of these amino acids; the other 10 must be obtained from the diet and are therefore known as essential amino acids. Three of these are the above-mentioned aromatic amino acids; animals lack the pathway for their synthesis including not having the EPSP synthase enzyme targeted by glyphosate. Thus glyphosate has minimal direct effects on animals. It thus has excellent toxicological properties, with numerous studies indicating it to be very safe. Glyphosate also rapidly loses its activity after use. It can be sprayed in a location one day and a day or two later highly susceptible plants can be planted in the same location with no effects. It also does not accumulate in the soil or waterways (http://www.speclab.com/compound/c1071836.htm).

As mentioned above, as of 2006, Roundup Ready crops along with Bt crops were the most widely grown biotech crops. Roundup Ready crops are resistant to Roundup and other glyphosate-based herbicides. Roundup Ready crops contain a form of the EPSP synthase isolated from the bacterium *Agrobacterium* sp. strain CP4 that is resistant to glyphosate. Use of Roundup Ready crops along with glyphosate-based herbicides such as Roundup can greatly facilitate weed control, especially in crops for which weed control has traditionally been challenging, such as soybeans. Glyphosate binds to the CP4 EPSP synthase in an unusual condensed conformation that is not inhibitory to the EPSP synthase reaction. A single-site mutation in the active site (Ala-100 to Gly) can restore glyphosate sensitivity to the

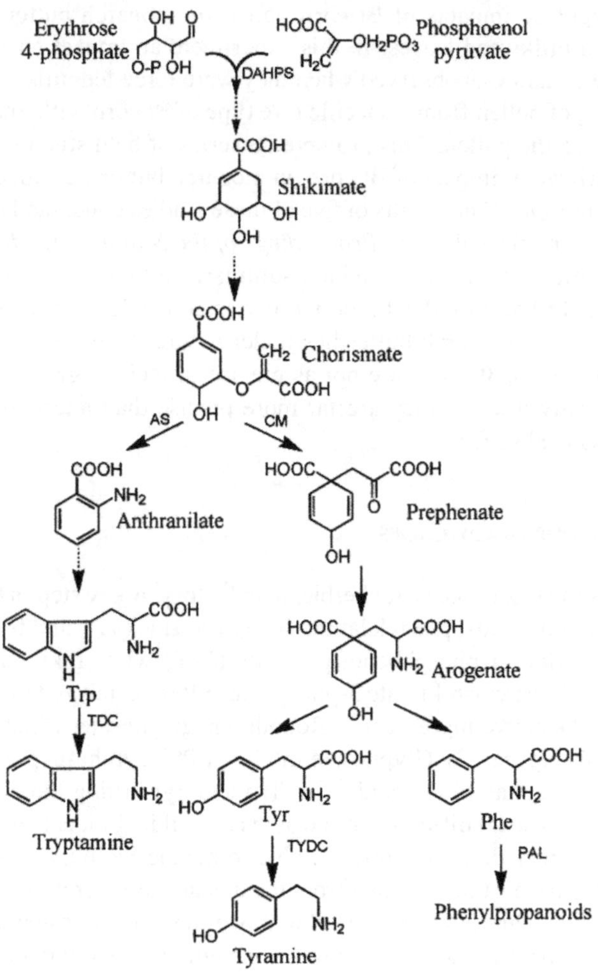

FIGURE 7.2 Biosynthesis of aromatic amino acids.

enzyme (Funke et al., 2006). A single point mutation changing proline-106 to leucine of a native plant EPSP synthase can also confer glyphosate resistance (Zhou et al., 2006). Roundup controls most weeds very effectively but there are a few, mainly dicot weeds, that are not effectively controlled by it. Such weeds can be effectively controlled by dicamba; dicamba-resistant crops have been developed and these are being combined with Roundup Ready crops (Behrens, 2003; Herman et al., 2005).

A major technological advance in the sustainable production of crops is conservation tillage or no-till crop production. This conserves topsoil and can reduce the fuel needed for crop production by 80% or more. The use of Roundup Ready or other herbicide-resistant crops greatly facilitates the use of conservation tillage (http://www2.ctic.purdue.edu/ctic/FINAL.pdf; http://www.omnistar.nl/brochures/OSInc_No_Till_Briefing_v2.pdf).

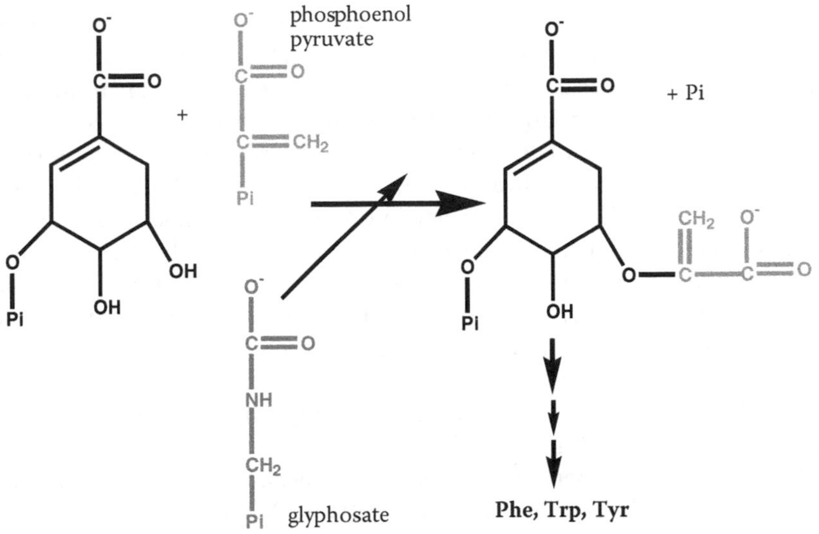

FIGURE 7.3 5-enoylpyruvylshikimate 3-phosphate (EPSP) is formed by condensation of phosphoenol pyruvate (PEP) with EPSP catalyzed by EPSP synthase. The active ingredient of Roundup is glyphosate, which is structurally similar to PEP. Glyphosate inhibits EPSP synthase preventing the formation of the essential aromatic amino acids phenylalanine (Phe), tryptophan (Trp) and tyrosine (Tyr) killing the plant tissues

7.3 NEW USES OF CROPS AND RENEWABLE RESOURCES

Most biotech crops grown by farmers to date, such as the Roundup Ready and Bt crops mentioned above, involve "input traits" or traits of value to farmers or growers. "Output traits," on the other hand, are traits that are of more interest to consumers. There is considerable current interest and effort to develop biotech crops for improved output traits, especially with respect to improved plant oils. The focus of new oilseed crops with new uses is on both improved human and animal nutrition and new industrial resources. Most plant cells and tissues contain five major common fatty acids, palmitic (16:0), stearic (18:0), oleic (18:1), linoleic (18:2) and α-linolenic (18:3) (Somerville et al., 2000). This is also true of most plant seed oils. Many plants, however, are known in nature to accumulate high levels of various unusual fatty acids in seed oils but not membrane lipids. Also, oilseeds are being genetically engineered to make fatty acids not normally made by higher plants.

7.4 IMPROVED HUMAN AND ANIMAL NUTRITION

Plant oils are generally healthier than animal fats because they are lower in saturated fatty acids and lack cholesterol. For many cooking applications, however, plant oils often have levels of polyunsaturated fatty acids that are too high for good oxidative stability, and traditionally, such oils have been partially hydrogenated for such cooking applications. This hydrogenation unfortunately results in the production of *trans*-fatty acids that are now well known to be undesirable in human health. Thus, there has been considerable effort to develop vegetable oils that are oxidatively stable without hydrogenation and therefore lack *trans* fatty acids resulting from hydrogenation. Oils with large increases in stearate contents with corresponding reductions in unsaturated fatty acids have been developed for stable cooking fats and oils without the need for hydrogenation (Knutzon et al., 1992; Kridl, 2002; Liu et al., 2002a, 2002b). Similarly, both breeding and genetic engineering have resulted in the development of high-oleate oils that are more oxidatively stable than regular oils for edible applications such as cooking, and for industrial applications such as for lubricants or biodiesel. For example, Kinney et al. at DuPont have developed soybeans with ~90% oleic acid in the seed oil simply by blocking the main Δ-12 desaturase, Fad2-1, responsible for converting seed 18:2 to 18:3 (Broun et al., 1999; Kinney, 1996). DuPont is collaborating with Bunge in marketing the high-oleate soybeans (Fyrwald and Hausmann, 2006).

One of the more important issues of diet and health is the dietary levels of ω-3 fatty acids and the balance of ω-3 and ω-6 fatty acids. The diets of most Westerners are deficient in ω-3 fatty acids and contain an excess of ω-6 fatty acids. The three main ω-3 fatty acids of nutritional significance are α-linolenic acid (18:3), eicosapentaenoic acid (EPA; 20:5) and docosahexaenoic acid (DHA; 22:6). Higher animals, including humans, cannot synthesize the ω-3 double bond but can convert α-linolenic acid into EPA and DHA, although not always efficiently.

All three of these ω-3 fatty acids are needed in healthy diets. Certain fish such as salmon, tuna and mackerel have traditionally been the main dietary sources. There is, however, a significant problem with using such fish as the main source of these important dietary components. First, global demand exceeds the level that can be acquired from wild catches. Second, fish can accumulate contaminants that threaten health such as mercury and polychlorinated biphenyls. Third, such fish do not actually make the ω-3 fatty acids, but simply accumulate them from the food chain. Farmed fish, such as salmon, need to have the ω-3 fatty acids in their diets for their accumulation of these fatty acids. It would be best and most sustainable to supply their ω-3 fatty acids from a crop source such as oilseeds with high levels of eicosapentaenoic and docosahexanoic acids.

Higher plants normally only make and accumulate α-linolenic acid, not the longer chain ω-3 fatty acids although many lower plants such as many marine algae make and accumulate eicosapentaenoic and docosahexaenoic acids. Some plants such as linseed or flax accumulate high levels (> 50%) of α-linolenic acid in their oil but no plants are known to have detectable levels of EPA and DHA in their seed oil. Recently considerable progress has been made in engineering oilseeds to make and accumulate these fatty acids in their seed oil (Kinney, 2006; Singh et al., 2005). It takes two additional desaturations and an elongation step to convert α-linolenic acid in plants into EPA and a third desaturation and another elongation for biosynthesis of DHA. Levels of eicosapentaenoic acid in seed oil similar to high accumulating fish have been achieved and development of high DHA accumulating oilseeds is in progress (Kinney, 2006). Progress is also being made in engineering oilseeds for increased vitamin E (tocopherol) levels, which will not only increase the nutritional value of such seeds but, also in combination with ω-3 fatty acid-accumulating oils, will help provide oxidative stability to such oils (Kinney, 2006).

7.5 NEW INDUSTRIAL RESOURCES

A very promising area of farming and agriculture in the future is use of plants as a source of renewable materials. Oilseeds are particularly attractive as a source of renewable specialty chemicals, and advances in biochemistry and molecular genetics are helping make this a reality. A pioneering achievement in engineering oilseeds as an enhanced source of industrial chemicals was the development of high-laurate (12:0) canola, for which laurate in the seed oil went from undetectable levels to ~60% (Voelker et al., 1996). Laurate is an important component of cold-water detergents.

Numerous other targets of oilseed genetic engineering for accumulation of industrially valuable fatty acids are being pursued, including oils that have high levels of acetylenic, conjugated, cyclopropyl, epoxy and hydroxy fatty acids (Table 7.2). These have applications as adhesives, coatings, composites, lubricants and resins. Genes encoding most of the enzymes catalyzing the biosynthesis of these fatty acids have been cloned and expressed in plants. In most cases, the substrate and products of these enzymes involves fatty acids esterified to phospholipids of the endoplasmic reticulum (ER). Plants are known in nature that can accumulate up to 90% of these unusual fatty acids, with this high accumulation exclusive to the oil or triacylglycerol (TAG) with levels in membrane lipids such as phospholipids generally maintained to only a few percent (Jaworski and Cahoon, 2003; Schmidt et al., 2006; Singh et al., 2005).

Genetic engineering of oilseeds for accumulation of these industrially valuable fatty acids have, with the exception of laurate canola mentioned above, generally resulted in 20% or less of them. High accumulators of these fatty acids clearly have a mechanism for the selective transfer into TAG. Candidates for this include phos-

TABLE 7.2
Some Targets of Oilseed Genetic Improvement

Oil Composition	Uses
High saturates	Shortenings and margarines
High oleate	Stable cooking oil, lubricants
High ω-3 oils	Nutrition, inks and coatings
Oils accumulating EPA and/or DHA[a]	Nutrition
High-laurate oils	Detergents
Epoxy fatty acid accumulating oils	Resins, coatings, adhesives
Hydroxy fatty acid accumulating oils	Lubricants, resins
Conjugated fatty acid accumulating oils	Coatings, resins
Acetylenic fatty acid accumulating oils	Antimicrobial agents, coatings
Cyclopropyl fatty acid accumulating oils	Lubricants

[a] EPA = eicosapentaenoic acid; DHA = docosahexaenoic acid

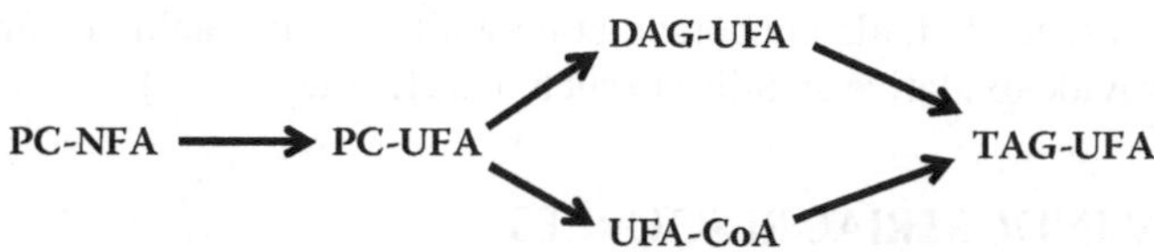

FIGURE 7.4 A normal fatty acid in phosphatidyl choline (PC-NFA) is converted into an unusual fatty acid still in PC (PC-UFA). The PC-UFA is in turn converted into diacylglycerol (DAG) and acyl CoAs containing such UFA and finally into triacylglycerol (TAG), the predominant component of oil.

pholipases, acyl CoA synthetases and final TAG biosynthetic enzymes with specificity for these specific fatty acids (Jaworski and Cahoon, 2003; Schmidt et al., 2006; Singh et al., 2005; Weselake, 2000; Weselake, 2001). The major ER membrane lipid, phosphatidyl choline (PC), is often the initial site for unusual fatty acid biosynthesis (Figure 7.4). Shockey et al. (2006) recently provided evidence that type 2 diacylglycerol acyltransferase (DGAT2) is particularly important in conjugated fatty acid accumulation in tung (*Vernicia fordii*) seed oil. This enzyme catalyzes the final step in seed oil formation.

7.6 BIOTECH CROPS: WHY THE CONTROVERSY?

Most everyone who has followed the applications of biotechnology in crop improvement understands that it has not been without controversy. As with most new technologies, crop genetic engineering can have possible associated risks along with benefits. That some would question what the risks are is expected and appropriate. The level of opposition that was reached in some circles, however, particularly in Western Europe in the late 1990s, was perplexing to many scientists working in the area. Some activists since the mid to late 90s have worked to frighten the public in order to minimize or prevent the use of recombinant DNA techniques in crop improvement. Curiously, in most countries of the world, GE plants are tightly regulated even though they contain similar genetic changes that scientists might be able to achieve by many other means, which are often are not regulated at all.

It is estimated that the regulatory costs for getting approval of a GE soybean line are between one and 10 million US$ (Schmidt et al., 2006). It is theoretically possible that genetic changes can be made to a native soybean EPSPS gene. These changes render it Roundup resistant by traditional mutagenesis, and breeding and modern molecular genetic techniques can be used to identify such mutants (Zhou et al., 2006). In many parts of the world, including the United States, such lines might not need any regulatory approval at all, which eliminates the associated regulatory costs. Thus, it might seem, with everything being equal, that a business would prefer to avoid genetic engineering of plants if the same trait can be achieved by other means.

From a business-development perspective, however, there are additional considerations. For a trait or crop that might be widely grown or utilized, genetic engineering could provide more robust intellectual property (IP) protection, possibly improving the return on the investment. IP issues are at the core of much of the controversy regarding GE crops. Antibiotech activists usually raise questions about food safety or environmental issues, because these are issues that the general public is usually more directly concerned with or knows more about than IP, and as such, these issues get a lot more attention. Studies indicate biotech crops are as safe or safer, however, than conventional crops and equal or better for the environment (Chassy et al., 2005). Because of this, other major issues that spur opposition among some antibiotech activists are IP and some seed business practices.

Traditionally, farmers have planted their own seeds, and shared or bartered seeds from neighboring farmers. Starting in the early part of the last century, production of hybrid seeds began to predominate in crops where hybrids show yield increases that more than cover the cost of the hybrid seeds. For example, in about the last 50 years, hybrid seed came to dominate corn production. The inbred parents are protected by the hybrid seed companies so farmers must buy the hybrid seed from seed companies every year and cannot replicate them for themselves. With self pollinated crops, where hybrid vigor is not very substantial, such as with soybeans, seed companies could do little to prevent farmers from planting their own seed, which limits the sales and profits of seed companies. Hybrid vigor is the situation wherein hybrids grow much more vigorously and yield significantly more than their inbred parents.

With the introduction of GE crops 11 years ago, seed companies selling these crops instituted a new business practice of having farmers sign legally binding documents stating that they will neither save their own seed nor provide the patented seeds to other farmers. Concurrently, there has been some consolidation in the seed industry, with many seed companies acquired by large chemical companies or companies that were predominately in chemical sales, especially, at least formerly, agricultural chemicals (Fulton and Giannakas, 2001; Kalaitzandonakes, 1998).

Some environmental groups have been traditionally at odds with chemical companies, so seeing much of the GE seed business ending up in the hands of such companies has helped fuel GE crop opposition. Some of these seed/biotech companies have been experimenting with technological means to prevent or minimize pirating of their seeds instead of using legal means. These are sometimes termed genetic use restriction technologies or GURTs (Lence and Hayes, 2005), and one such GURT that had been pursued by a major biotech seed company had been called "terminator seeds." Such seeds can be planted like regular seeds and produce grain (seed) yields like regular crops that can be processed into foods, feed etc. as usual, but the progeny seed do not germinate, resulting in the farmer having to go back to the source and buy the seed every year.

Such seeds generated considerable controversy when their utilization was first considered by a major seed company in the mid to late 1990s. The technology was not implemented by the original company that investigated it, perhaps due in part to the controversy it created. This is akin to software and media companies trying to prevent copying of CDs and DVDs, and pharmaceutical companies trying to prevent generic production of drugs that are still under patent. Patents have been widely considered by businesspeople and economists to be an important stimulator of technological and economic developments, but social scientists have traditionally had mixed opinions about patents. The concerns include the fear that wealthier individuals and companies might be better positioned to gain patents, and patented products might not be affordable by lower income individuals. In other words, business/investor interests can be at odds with consumer interests and patents might exacerbate this. In the case of seed or food crops it can be argued that this is a more serious issue than others because food is necessary for our existence and farming traditions are considered more sacred than other businesses.

A related concern is if some biotech crops become so popular with farmers that some of their seeds handed down by their parents that they traditionally planted are not maintained, the genetic diversity of these crops could be lost. Several organizations of many governments, such as the United States Department of Agriculture (USDA), maintain large germplasm collections of many crops to preserve such genetic diversity. There is a movement to maintain germplasm collections from many parts of the world, including countries that lack resources to maintain germplasm collections themselves, in an international seed bank in northern Norway (Charles, 2006).

A concern that some have voiced, which is related to IP issues and the seed business, is whether crop biotechnology favors large companies at the expense of smaller companies (Lesser, 1998). Unlike machinery, where the economy of scale usually favors large producers at the expense of smaller farmers, technologies in seeds themselves are thought to be scale-neutral. Larger, wealthier farmers, however, may learn about or have access to improved seeds earlier than smaller, poorer producers. Also, larger companies can often have more extensive research and development groups. Government regulatory hurdles to bringing GE crops to market are one of the safeguards pushed by antibiotech activists. Oddly enough, this can favor large companies at the expense of smaller seed companies because large companies can afford to do the work to meet the regulatory requirements and can maintain departments that deal effectively with regulatory issues (Miller and Conko, 2004).

Traditionally, improved crop cultivars that have been grown by farmers have been developed by university or government scientists and made freely available to the public. The U.S. Plant Patent Act of 1930 first allowed the patenting of asexually propagated plant varieties (or cultivars) with the exception of tubers. In the 1960s, some European countries afforded breeders IP rights to sexually reproduced crop cultivars. The Plant Variety Protection Act of 1970 in the U.S. also provides IP protection to novel plant cultivars for 18 years (http://www.nalusda.gov/pgdic/Probe/v2n2/plant.html). A number of governments moved to further privatize this process in the 1980s. In the United States, for example, federal support for plant breeding has disappeared, and the federal government's policy is that this is the responsibility of the private sector. The USDA, for example, up until the 1980s, employed many plant breeders and was responsible for developing many crop cultivars grown extensively by farmers up until this period ~20 years ago. It no longer employs plant breeders, however, and does not release crop cultivars anymore. In the ensuing period, cultivar development of major crop plants has come to be dominated by the private sector, much like engineering products and business have been for a long time.

Since the introduction of biotech crops, corporate involvement and IP ownership in crop seeds has expanded. Additionally, many scientific discoveries made by university and government scientists that have practical utility are often patented and licensed to private companies. Naturally, there are those who do not like this change in tradition, which is not universally supported by the agricultural community. There is a concern that resource-limited farmers might not have access to such developments or scientists might be limited in their freedom to apply such technological improvements to minor crops. A group known as the Public Intellectual Property Resource for Agriculture, or PIPRA, has been formed to help address these issues. PIPRA is an initiative by universities, foundations and nonprofit research institutions to make agricultural technologies more easily available for development and distribution of subsistence crops for humanitarian purposes in the developing world and specialty crops in the developed world (http://www.pipra.org/). Additionally, the IP Modeling Group is trying to understand how to calibrate IP and patent law policy so as to maximize benefits to society (http://www.cipp.mcgill.ca/en/projects/ipmg/).

An additionally complicating factor in the biotech crop controversy is the backdrop of agricultural trade disputes that have been going on for several decades. This has led to a major divide between exporting and importing countries of major agricultural commodities. Major exporting countries of canola, corn, rice, soybeans and wheat, such as Canada, South American countries, South Africa and Australia, have extensively adopted biotech crops, and major importing countries such as Western Europe and Japan have not. Exporting countries have accused importing countries of using restrictions on biotech crops as "non-tariff trade barriers." Although this may be true, there are many other factors to be considered (Wilson et al., 2001).

Many Western European countries and Japan want to maintain many of their small family farms, and biotechnology is seen as yet another technology that makes their farms less competitive on the global market. There is even a debate in Britain as to whether the government should continue to subsidize farms or whether they should instead employ many of the farmers as wild-habitat managers. Some ecologists argue that maintaining land in agricultural uses needlessly is not as ecologically desirable as converting it back to wild habitat. Much of the land in the northeast U.S. was farmed by early settlers, and since then much of it in rural areas has been diverted back into wild habitat as more productive land in other parts of the U.S. became farmed, and scientific and technological advancements greatly improved crop yields per unit land area. This, of course, has not been without social and economic consequences, and many in the United States, who once derived a living from agriculture, now need to make a living in different ways.

Also, with fewer farmers, there can be a ripple effect in rural economies. Although many technological advances have contributed to this process, the applications of biotechnology to developing new uses of plants can provide new opportunities for farmers and rural communities. A very important potential contribution of science and technology to crop improvement will be to increase yields per unit land area, thereby helping stem wild habitat destruction in many developing areas of the world such as tropical rain forests (Chassy et al., 2005; Jauhar, 2006).

7.7 EDUCATING THE PUBLIC ON PLANT LIPID BIOTECHNOLOGY

Fundamental to educating the public on plant lipid biotechnology is improving the general knowledge of plant breeding and genetics/DNA science. It is also important to dispel major misconceptions associated with biotech crops. The term genetically modified organism (GMO) has been so thoroughly associated with some misconceptions that it may be best to avoid using this term when possible. Historically, few, except scientists involved in breeding and genetics, have had much of an understanding or interest in how the crops grown by farmers were developed. A question on a survey conducted in many nations in recent years asked: only GMO crop derived foods contain DNA, true or false? Well over 10% of the respondents said this was true. Furthermore, it is not uncommon for the lay public to be concerned about what effect consuming GMO corn might have on their own DNA. The good news about this is that most people now understand that they themselves have DNA and the GMO controversy has increased interest in the subject, although this interest may be waning as GMOs are less frequently in the news.

It is obvious, however, that few have considered the genetics of the crop plants they consume. In fact, it is common for many nonscientists to be unclear about whether plants are actually living organisms, as most plants do not move with sufficient speed to be readily observed as movement (and we naturally associate motility with life). Thus, a little education on general biology is also useful. Interestingly, although surveys indicate the public finds the term GMO to be frightening and negative, such surveys indicate they have a positive attitude about biotechnology.

Outlined below are some fundamental concepts that we attempt to transmit to students via our introductory course in DNA Science/Crop Biotech at the University of Kentucky.

7.7.1 DNA Science/Crop Biotech 101

- Plants, bacteria, fungi as well as animals are all living organisms.
- All living organisms have genes that encode the cumulative traits or phenotype that make up that organism.
- The phenotype of all organisms, just like for you and me, is a consequence of our genotype and environment.
- Genetic information is coded in DNA of chromosomes. The DNA code can be considered a "genetic alphabet."
- There are four letters (bases) in the DNA code, A, C, G and T. These DNA bases are precisely the same in all living organisms.
- The DNA bases are to an organism as letters of written alphabets are to books; genes are like words composed of different letters, traits are like sentences and whole organisms like books.
- Crop plants, including oilseeds, produced by traditional breeding have many genes (words) changed relative to their parents, and GE crops have fewer words changed but the changes are more directed and precise. The DNA building blocks (bases or letters) are the same.

- Most foods we consume contain DNA, although in oilseed refining the DNA is retained in the meal, not extracted into the oil. DNA in our foods is digested just like protein when consumed. After digestion, DNA from all living sources, GMO or not, is exactly the same.

Scientists working in breeding and genetics need to increase and improve their communication of these principles to the general public. The busy lives and competing interests of most adults are such that most will not pay much attention to such efforts but those with concerns about our food and how it is produced might. In the long run, DNA science and genetics need to be incorporated into general science education throughout the world. This is happening, but not as fast or uniformly as would be ideal, and the connections to biotech crops are not always apparent. The rationale for this is that our health is dependent on our own genetic information or genotype along with our environment/life style.

Also, DNA science and genetics are revolutionizing medicine and medical applications, including new treatments and pharmaceutical developments. Similarly, DNA science and biotech are revolutionizing agriculture and various applications of crop products including oils. As mentioned earlier, much healthier oils are being developed via biotechnology. Also, DNA science/biotech can contribute to more environmentally friendly and sustainable crop production. Moreover, genetic engineering, together with chemistry and engineering advancements, will contribute enormously to future renewable resource needs from fuels for production of numerous useful chemicals and materials. Working with science teacher associations, such as the U.S. NSTA (http://www.nsta.org/index.html), can be useful in this regard. Iowa Extension has developed Biotechnology Curriculum Units for different grade levels: http://www.biotech.iastate.edu/publications/ed_resources/biotech_curriculum.html.

Questions on DNA science and genetics need to be incorporated into national assessment and college entrance exams to encourage teachers, students and parents to learn more about this important subject. Access Excellence at the National Health Museum has developed a unit on food biotechnology for high school teachers and students called "Dining on DNA" that is available online: http://www.accessexcellence.org/RC/AB/BA/DODpub/. An animated primer on the basics of DNA, genetics and heredity with a medical focus is available at http://www.dnaftb.org/dnaftb/.

Introducing DNA science and biotech to youth groups is another useful way to educate the public. Various 4-H organizations, for example, have theme activities and camps on DNA science/genetics, often using a crime scene investigation (CSI) theme, since this appeals to nonscientists as well as the scientifically inclined (http://www.ces.ncsu.edu/depts/agcomm/writing/2004/biotechcamp.htm; http://www.ca.uky.edu/brei/4-H_Program.htm). A good resource for youth leaders with little knowledge of biotechnology, developed by Iowa Cooperative Extension, "*A Crime, A Clue, and Biotechnology,*" is available for a nominal fee (~$20 including shipping) from Iowa Cooperative Extension: http://www.extension.iastate.edu/e-set/biotech.html.

In 1997, the U.S. National 4-H Council's Environmental Stewardship program published the excellent leader's guide called "Making Sense of Biotechnology in

Agriculture: Fields of Genes." It is available at http://fog.n4h.org/. The Wisconsin BioTrek program has well-developed biotech educational outreach activities for the public, http://www.biotech.wisc.edu/outreach/outreach.html, and is willing to help with similar outreach efforts in other parts of the world. An effective way to introduce DNA science/biotech to kids of all ages is to have them extract and look at DNA from some common fruit or vegetable and then make a smoothie with the same fruit or vegetable and talk about genetics and DNA as they consume it. It works best with something they like to eat such as bananas or strawberries. Protocols are available on the web for extracting and spooling DNA from common foods, such as bananas, using items available at a regular grocery store (http://apps.caes.uga.edu/sbof/main/index.cfm?page=liatpics&trunkName=Banana%20DNA%20Extraction, http://kvhs.nbed.nb.ca/gallant/biology/dna_banana.html, http://kvhs.nbed.nb.ca/gallant/biology/dna_banana.html and http://www.ca.uky.edu/brei/Teach/4-H/Info/DNA%20Sci%20Exercises.htm). Some of these also include DNA science/biotech lesson plans.

ACKNOWLEDGMENTS

The author thanks Hirotada Fukushige and Tara Burke for helpful editorial comments.

REFERENCES

Behrens, M.R. 2003. Genetic engineering of tobacco and tomato plants for resistance to the herbicide dicamba. M.S., University of Nebraska, Lincoln.

Broun, P., S. Gettner, and C. Somerville. 1999. Genetic engineering of plant lipids. *Annu. Rev. Nutr.* 19:197–216.

Charles, D. 2006. Species conservation: A 'forever' seed bank takes root in the Arctic. *Science* 312:1730b–1731.

Chassy, B.M., W.A. Parrott, and R. Roush. 2005. Crop biotechnology and the future of food: A scientific assessment. CAST commentary, http://www.dast-science.org/websiteuploads/publicationspdfs/QTA2005-2.pdf.

Doebley, J. 2006. Plant science: Unfallen grains: How ancient farmers turned weeds into crops. *Science* 312:1318–1319.

Fernandez, H., S. Hughes, J.-D. Vigne, D. Helmer, G. Hodgins, C. Miquel, C. Hanni, G. Luikart, and P. Taberlet. 2006. Divergent mtDNA lineages of goats in an Early Neolithic site, far from the initial domestication areas. Proceedings of the National Academy of Science (PNAS) 103:15375–15379.

Fulton, M., and K. Giannakas. 2001. Agricultural biotechnology and industry structure. *AgBioForum* 4:137–151.

Funke, T., H. Han, M.L. Healy-Fried, M. Fischer, and E. Schonbrunn. 2006. Molecular basis for the herbicide resistance of Roundup Ready crops. PNAS 103:13010–13015.

Fyrwald, E., and C. Hausmann. 2006. DuPont and Bunge broaden soy collaboration to include industrial applications and biofuels, http://vocuspr.vocus.com/VocusPR30/Newsroom/Query.aspx?SiteName=DupontNew&Entity=PRAsset&SF_PRAsset_PRAssetID_EQ=102994&XSL=PressRelease&Cache=False (verified September 6, 2006).

Hanotte, O., D.G. Bradley, J.W. Ochieng, Y. Verjee, E.W. Hill, and J.E.O. Rege. 2002. African pastoralism: Genetic imprints of origins and migrations. *Science* 296:336–339.

Hartl, D.L., and E.W. Jones. 2005. *Genetics: Analysis of Genes and Genomes*, 6th Edition. Jones and Bartlett. Sudbury, MA.

Herman, P.L., M. Behrens, S. Chakraborty, B.M. Chrastil, J. Barycki, and D.P. Weeks. 2005. A three-component dicamba O-demethylase from Pseudomonas maltophilia, strain DI-6: Gene isolation, characterization, and heterologous expression. *J. Biol. Chem.* 280:24759–24767.

Horsch, R.B., J.E. Fry, N.L. Hoffman, D. Eichholtz, S.G. Rogers, and R.T. Fraley. 1985. A simple and general method for transferring genes into plants. *Science* 227:1229–1231.

Jauhar, P.P. 2006. Modern biotechnology as an integral supplement to conventional plant breeding: The prospects and challenges. *Crop. Sci.* 46:1841–1859.

Jaworski, J., and E.B. Cahoon. 2003. Industrial oils from transgenic plants. *Curr. Opin. Plant Biol.* 6:178–184.

Kalaitzandonakes, N. 1998. Biotechnology and the restructuring of the agricultural supply chain. *AgBioForum* 1:40–42.

Kinney, A. 1996. Development of genetically engineered soybean oils for food applications. *J. Food Lipids* 3:273–292.

Kinney, A.J. 2006. Metabolic engineering in plants for human health and nutrition. *Curr. Opin. Biotechnol.* 17:130–138.

Knutzon, D.S., G.A. Thompson, S.E. Radke, W.B. Johnson, V.C. Knauf, and K.J. C. 1992. Modification of Brassica seed oil by antisense expression of a stearoyl–acyl carrier protein desaturase gene. Proceedings of the National Academy of Sciences of the United States of America 89:2624–2628.

Konishi, S., T. Izawa, S.Y. Lin, K. Ebana, Y. Fukuta, T. Sasaki, and M. Yano. 2006. An SNP Caused loss of seed shattering during rice domestication. *Science* 312:1392–1396.

Kridl, J. 2002. Method for increasing stearate content in Soybean Oil Patent. 6,380,462 2002.

Larson, G., K. Dobney, U. Albarella, M. Fang, E. Matisoo-Smith, J. Robins, S. Lowden, H. Finlayson, T. Brand, E. Willerslev, P. Rowley-Conwy, L. Andersson, and A. Cooper. 2005. Worldwide phylogeography of wild boar reveals multiple centers of pig domestication. Science 307:1618–1621.

Lence, S.H., and D.J. Hayes. 2005. Technology fees versus GURTs in the presence of spillovers: World welfare impacts. *Agbioforum* 8:2–3.

Lesser, W. 1998. Intellectual property rights and concentration in agricultural biotechnology. *AgBioForum* 1:56–61.

Li, C., A. Zhou, and T. Sang. 2006. Rice domestication by reducing shattering. *Science* 311:1936–1939.

Liu, Q., S.P. Singh, and A.G. Green. 2002a. High-oleic and high-stearic cottonseed oils: Nutritionally improved cooking oils developed using gene silencing. *J. Am. Coll. Nutr.* 21:205S–211S.

Liu, Q., S.P. Singh, and A.G. Green. 2002b. High-stearic and high-oleic cottonseed oils produced by hairpin RNA-mediated post-transcriptional gene silencing. *Plant Physiol.* 129:1732–1743.

Miller, H.-I., and G.-P. Conko. 2004. *The Frankenfood myth: How protest and politics threaten the biotech revolution,* Praeger, Westport, CT.

Palmiter, R.D., G. Norstedt, R.E. Gelinas, R.E. Hammer, and R.L. Brinster. 1983. Metallothionein–human GH fusion genes stimulate growth of mice. *Science* 222:809–814.

Savolainen, P., Y.-P. Zhang, J. Luo, J. Lundeberg, and T. Leitner. 2002. Genetic evidence for an East Asian origin of domestic dogs. *Science* 298:1610–1613.

Schmidt, M.A., C.R. Dietrich, and E.B. Cahoon. 2006. Biotechnological enhancement of soybean oil for lubricant applications, pp. 389–397. *In* L. R. Rudnick, ed. *Synthetics, Mineral Oils, and Bio-Based Lubricants*, Chemistry and Technology Series, Vol. 23. CRC/Taylor & Francis Group, Boca Raton, FL.

Shockey, J.M., S.K. Gidda, D.C. Chapital, J.-C. Kuan, P.K. Dhanoa, J.M. Bland, S.J. Rothstein, R.T. Mullen, and J.M. Dyer. 2006. Tung tree DGAT1 and DGAT2 have nonredundant functions in triacylglycerol biosynthesis and are localized to different subdomains of the endoplasmic reticulum. *Plant Cell* 18:2294–2313.

Singh, S.P., X.-R. Zhou, Q. Liu, S. Stymne, and A.G. Green. 2005. Metabolic engineering of new fatty acids in plants. *Curr. Opin. Plant Biol.* 8:197–203.

Somerville, C., J. Browse, J. Jaworski, and J. Ohlrogge. 2000. Lipids, pp. 456–527. In B. Buchanan, et al., eds. *Biochemistry and Molecular Biology of Plants.* American Society of Plant Physiologists, Rockville, MD.

Voelker, T.A., T.R. Hayes, A.M. Cranmer, J.C. Turner, and H.M. Davies. 1996. Genetic engineering of a quantitative trait: Metabolic and genetic parameters influencing the accumulation of laurate in rapeseed. *Plant-Journal* (United Kingdom) 9:229–241.

Weiss, E., M.E. Kislev, and A. Hartmann. 2006. ANTHROPOLOGY: Autonomous cultivation before domestication. *Science* 312:1608–1610.

Weselake, R. 2000. The slippery roads to oil and fat. *Inform* 11:1281–1286.

Weselake, R.J. 2001. Biochemistry and biotechnology of triacylglycerol accumulation in plants, pp. 27–56. In T.M. Kuo and H.W. Gardner, eds. *Lipid Biotechnology.* Marcel Dekker, Inc., New York.

Wilson, R.T., C.T. Hou, and D.F. Hildebrand. 2001. *Dealing with Genetically Modified Crops*, AOCS Press, Champaign, IL.

Zambryski, P., M. Holsters, K. Kruger, A. Depicker, J. Schell, M. Van Montagu, and H.M. Goodman. 1980. Tumor DNA structure in plant cells transformed by *A. tumefaciens. Science* 209:1385–1391.

Zhou, M., H. Xu, X. Wei, Z. Ye, L. Wei, W. Gong, Y. Wang, and Z. Zhu. 2006. Identification of a glyphosate-resistant mutant of rice 5-enolpyruvylshikimate 3-phosphate synthase using a directed evolution strategy. *Plant Physiol.* 140:184–195.

8 Understanding Isoprenoid Biochemistry

Mee-Len Chye

CONTENTS

8.1 WHAT ARE ISOPRENOIDS?

Isoprenoids make up a group of hydrocarbons found in unicellular and multicellular organisms (Eubacteria, Archaebacteria, protozoa, fungi, algae, insects, animals and plants) that are diverse in chemistry, structure and function.[1–3] Over 30,000 natural isoprenoid compounds have been characterized and still more are being discovered and identified. The precursors of isoprenoids are isopentenyl diphosphate (IPP) and dimethylallyl diphosphate (DMAPP), isomers which constitute the building blocks for the formation of this large family of naturally occurring compounds.[1,4] Isoprenoid compounds that are generated from the C5 monomers IPP and DMAPP include C10 monoterpenes, C15 sesquiterpenes, C20 diterpenes, C30 triterpenes and C40 tetraterpenes.[1,4]

8.2 WHERE ARE ISOPRENOIDS FOUND?

Many diverse isoprenoid compounds occur naturally in the plant kingd

ese compounds, e.g., menthol, limonene, camphor, geraniol. Plant resins contain terpenes, from which rosin and turpentine are obtained. Other plant-derived isoprenoids include terpenes (linalool, linalyl acetate, terpinen-4-ol, citronellol and α-pinene),

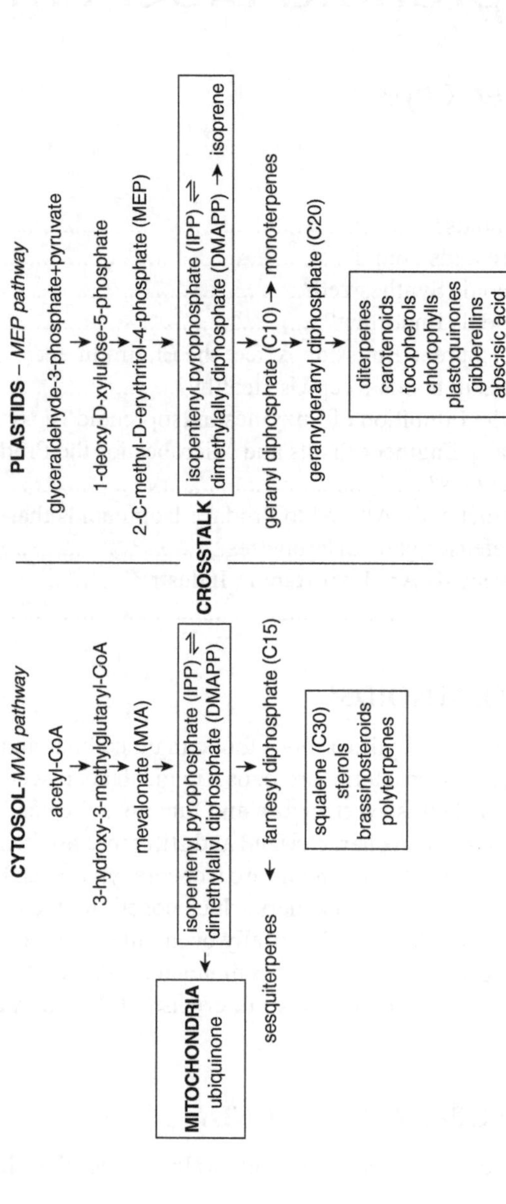

FIGURE 8.1 Pathways in isoprenoid biosynthesis. Comparison of isoprenoid production via the mevalonate (MVA) and the non-mevalonate (MEP) pathways in an organism in which both are present, e.g., plants. Crosstalk has been reported between these independent pathways.

which are utilized as cutaneously applied essential oils.[5] Natural rubber, guayule and gutta-percha are polyterpenes. The plant hormones abscisic acid and gibberellic acid, are also isoprenoid derivatives. Sterols that are biologically important in the formation of membrane structures and hormones are C30 isoprenoid compounds.[4] Other examples include C40 carotenoids, chlorophylls and ubiquinones, which are composed of isoprenoid chains (Figure 8.1).

In mammals, examples of isoprenoid-containing compounds are cholesterol, oxysterols, steroid hormones, bile acids, retinoids, heme A, ubiquinones, dolichols, prenylated proteins and isopentenylated tRNAs.[6] Cholesterol is an important constituent of mammalian membranes, while its oxidized derivative, oxysterol, plays a significant role in molecular signaling.[7,8] Other isoprenoid-derived bioactive molecules are involved in signaling and in gene regulation; thus they regulate many cellular events in eukaryotes.[6,8] Other examples of isoprenoid compounds include juvenile hormone and ecdysone in insects, prenylated mating hormones in fungi and ergosterols in yeast.[6] Membrane lipids from Archaebacteria and Eubacteria also consist of isoprenoid derivatives.

8.3 HOW ARE ISOPRENOIDS SYNTHESIZED?

Isoprenoids are synthesized via the classical mevalonate (MVA) pathway or the recently discovered non-mevalonate methylerythritol phosphate (MEP or Rohmer) pathway.[9–11] The MEP pathway occurs in phototrophs (algae and higher plants) and in most Eubacteria.[11,12] In Archaebacteria, some Eubacteria and eukaryotes, the MVA pathway is utilized to generate IPP.[2] In some eukaryotes in which both pathways coexist, MVA-associated isoprenoid biosynthesis occurs in the cytosol while the MEP pathway operates in plastids. In general, non-plastidic eukaryotes synthesize isoprenoids via the MVA pathway alone while both pathways coexist in higher plants, some algae, mosses and liverworts, and marine diatoms.[2] Interestingly, both pathways are used to form IPP in the bacterium *Streptomyces*.[12] There is increasing evidence of crosstalk between these two independent pathways that regulate isoprenoid biosynthesis in organisms in which both coexist.[13–16]

The MVA pathway generates isoprenoids targeted for the biosyntheses of sterols, brassinosteroids, triterpenes, sesquiterpenes, polyterpenes, squalene, dolichol, ubiquinones, oxysterols, farnesol, geranyl geranyl and prenylated proteins.[4,6,9] This pathway is vital to animal and plant cells, and many end products arising from MVA biosynthesis are necessary for growth and development.[6,17,18] MVA, the precursor to cytosolic isoprenoid compounds, is the product of three enzymes (acetoacetyl-CoA thiolase, 3-hydroxy-3-methylglutaryl-CoA [HMG-CoA] synthase and HMG-CoA reductase) by which acetyl-CoA is converted to HMG-CoA and eventually to MVA.[19] Subsequently, MVA is converted to IPP, the building block for synthesis of further isoprenoids (Figure 8.1).

The MEP pathway is closely associated with photosynthesis, which supplies the carbon source as well as the required reducing electron flow for this pathway.[20] This plastidial pathway generates volatile isoprenoids (isoprene C_5H_8, monoterpenes), diterpenes, tetraterpenes and isoprenoids for the biosyntheses of photosynthetic pigments (carotenoids and the side chains of chlorophylls, tocopherols [vitamin E], phylloquinones and plastiquinones) and phytohormones including abscisic acid, cytokinins and gibberellins.[3,4,15] In the MEP pathway,[2] condensation of pyruvate and glyceraldehyde-3-phosphate by DXP synthase results in the formation of 1-deoxyxylulose-5-phosphate (DXP). The next enzyme deoxyxylulose phosphate reducto-isomerase (DXR) converts DXP to 2-C-methyl-D-erythritol-4-phosphate, which eventually yields IPP/DMAPP (Figure 8.1).

8.4 WHY ARE ISOPRENOIDS IMPORTANT?

In animals, cholesterol is derived from IPP, while in yeast, sterols are produced. Because isoprenoids function as membrane structures, reproductive hormones, mating pheromes, photoprotective agents, visual pigments and signal transduction components, defects in isoprenoid production have been implicated in cancer and coronary artery disease.[1,8] Cholesterol has an important role in the regulation of lipid metabolism, and mammalian cellular cholesterol levels are maintained within a narrow range,[7] deviations from which may give rise to lipid-associated diseases in humans, including atherosclerosis, Tangier disease and Niemann-Pick disease type C. The administration of cholesterol-lowering agents has been shown to delay the onset of Alzheimer's disease.[7]

Many important medicinal plant compounds are derived from isoprenoid biosynthesis. Taxol from *Taxus* species has anti-cancer properties, while the anti-malarial drug artemisinim is obtained from *Artemisia annua*.[21] Limonene and perillyl alcohol are monoterpenes potentially useful as anti-cancer therapeutic agents.[22] *Schisandra* triterpenoids display anti-HIV-1 activity[23] while *Sagittaria* diterpenoids possess antibacterial activity.[24] Limonoids, which give the bitter taste of citrus, are triterpenes with anti-microbial, anti-malarial, anti-cancer and insecticidal properties.[25]

Pollination, an important process in the life cycle of a plant, relies on floral scent production that is due to the emission of terpenoids and benzenoids.[26] In snapdragon flowers, it has been reported that these floral-derived terpenoids are produced via the MEP pathway.[26]

There are some 600 carotenoids distributed across species that constitute the accessory photosynthetic pigments in phototrophs.[27] Some of these caroteinoids function in photoprotection by the removal of excess energy generated during photosynthesis or detoxification of reactive oxygen species.[3] This ability to physically deactivate reactive oxygen species is due to the presence of the double bonds that are found in isoprenoid compounds.[3] Furthermore, endogenous isoprene production has an antioxidant function in plants, protecting the isoprene-emitting plant against ozone damage.[28] The production of isoprene and the monoterpenes, limonene and pinene, is beneficial to the plant in conferring the ability to tolerate high temperatures in the light.[3]

8.5 HOW DOES PLANT ISOPRENE EMISSION AFFECT THE ENVIRONMENT?

Up to 10% of carbon fixed during photosynthesis can be lost by emission to the environment as phytogenic volatile organic compounds (VOCs), which include alkanes, alkenes, alcohols, aldehydes, ethers, esters and carboxylic acids, with the most abundant group being isoprenoid in nature.[4,29] Such VOC emission is due to the volatility of these compounds. It has been suggested that emissions function as metabolic safety valves in removal of excess energy or carbon that accumulates upon stress, in physicochemical protection that stabilizes membranes against oxidative stress, and in plant–plant, plant–insect or plant–pathogen communication in defense.[29]

There are many reasons as to why some vascular and nonvascular plants emit isoprene that is synthesized in chloroplasts from DMAPP by isoprene synthase.[3,30,31] Such isoprene emissions are believed to help plants survive rapid alterations in air temperature and to protect photosynthetic membranes from thermal damage. Given that the amount of isoprene released by plants to the atmosphere is comparable to that of methane, and that isoprene is reactive to hydroxyl radicals, this substantial emission of isoprene will no doubt affect the oxidizing potential.[30,31]

The rate of isoprene emission has been reported to be correlated with elevated temperature and ozone levels.[30] Light-dependent isoprene production is induced by water stress and is inhibited under conditions of darkness and by inhibitors of photosynthesis.[30] Hence, it is believed that isoprene production may actually have an evolutionary significance in permitting plants to tolerate and survive rapid fluctuations in temperature by the ability of isoprene to dissolve into membranes, thus aiding the plant to withstand high temperature stress.[30,31] Nonetheless, the presence of isoprene in the atmosphere has an adverse effect on the environment due to its oxidation, culminating in the production of ozone, blue haze and smog in the presence of nitrogen oxides.[31]

8.6 WHICH ISOPRENOIDS IN FOODS KEEP US HEALTHY?

Many isoprenoid compounds found naturally in foods have been demonstrated to benefit human health. These include plant sterols, carotenoids and triterpenoid derivatives. Plant sterols (phytosterols) have been tested in trials involving their application as functional foods because they promote the lowering of blood cholesterol levels and hence can be used to combat atherosclerosis and cardiovascular disease in the human population.[32,33] Previously, both the intraperitoneal and subcutaneous administration of phytosterol had been reported to reduce plasma cholesterol levels.[32] Specifically, phytosterol intake in the diet as enriched margarines has been demonstrated to reduce cholesterol levels in humans, and it has been proposed that such incorporation of phytosterols in the diet will bring about other beneficial effects.[34] Phytosterols are believed to inhibit cholesterol uptake in the gut.[32] Furthermore, the ingestion of plant sterols has been demonstrated to impede the development of colon cancer.[32]

Carotenoids display antioxidant activities and are thus significant in promoting human health, in particular immune function, while the provitamin A carotenoids are important for vision.[22] Astaxanthins are carotenoids that have been suggested to enhance human health due to their antioxidant and anticancer properties.[35] Also, it has been demonstrated that the dietary triterpenic compounds (olealonic acid, maslinic acids, erythrodiol and uvol) from pomace olive oil can induce vasorelaxation of the aortas in spontaneously hypertensive rats.[36]

8.7 APPLICATIONS BY THE INHIBITION OF ENZYMES IN ISOPRENOID BIOSYNTHESIS

Mevinolin is a fungal metabolic from which mevinolinic acid is derived. Mevinolinic acid is an inhibitor of HMG-CoA reductase, an enzyme in the MVA pathway, and it has been used as a drug to lower plasma cholesterol because HMG-CoA reductase is a rate-limiting step in cholesterol biosynthesis.[37] More than half of the total body cholesterol in humans is synthesized *de novo,* and therefore HMG-CoA reductase is an appropriate target for cholesterol-lowering agents.

The antibiotic fosmidomycin specifically inhibits deoxyxylulose phosphate reducto-isomerase (DXR), the second enzyme in the MEP pathway.[38] Fosidomycin and other inhibitors such as phosphonohydroxamic acids have demonstrated potential as antibacterial and antiparasitic drugs, targeting isoprenoid biosynthesis in pathogenic bacteria and in *Plasmodium falciparum* (the causative agent for malaria), which has evolutionary links to photosynthetic organisms.[39,40] Inhibition of the MEP pathway is detrimental to these disease-causing organisms because isoprene biosynthesis is essential for their viability.

For those organisms in which both MVA and MEP pathways are present, the picture is more complex when only one such pathway is inhibited, as crosstalk has been reported to occur between these independent pathways. For example, the development of herbicides by using inhibitors that target the MEP pathway to block the biosynthesis of chlorophyll, carotenoids and phytohormones will not be feasible if the MVA pathway can compensate for the loss arising from the interruption of the MEP pathway.[16]

Recently, the MVA pathway has been targeted for anti-cancer therapy by the inhibition of HMG-CoA reductase or farnesyl prenyl transferase and geranyl geranyl prenyl transferase, which are implicated in the posttranslational modification of proteins like small GTPases including Ras and Ras-related proteins.[41] Such inhibition, brought about by the administration of drugs that block the MVA pathway or those that affect protein prenylation, have altered the growth and survival of malignant cells in cultures and in animal models, demonstrating potential for further use in the treatment of human cancers.[41] Despite success in preclinical investigations, conducting clinical trials has proven more challenging than anticipated because of potential problems associated with the clinical use of these drugs.[41]

8.8 CAN WE GENETICALLY ENGINEER PLANTS AND MICROBES FOR THE PRODUCTION OF DESIRED ISOPRENOIDS?

Because of the beneficial health effects shown by the incorporation of phytosterols into our diets, scientists have worked toward the possibilities in the overproduction of health-promoting isoprenoids, such as phytosterols, in transgenic crops. Demonstrations of the benefits of such genetic manipulations are usually first tested out in model plants such as tobacco and Arabidopsis before these strategies can be applied to crop plants, because crop species are usually more difficult to genetically transform.

In initial experiments toward this end, the enzyme tested in the genetic manipulation of phytosterol biosynthesis was HMG-CoA reductase, a rate-limiting enzyme of the MVA pathway. It was demonstrated that the overexpression of *hmg1*[42] encoding *Hevea brasiliensis* HMG-CoA reductase increased sterol content by 6-fold in leaf tissue of transgenic tobacco.[43] Further, the expression of a truncated version of *Hevea* HMG-CoA reductase enhanced total seed sterol content by 3.2-fold in transgenic tobacco,[44] while the expression of a truncated version of hamster HMG-CoA reductase increased total leaf sterol by 3- to 10-fold.[45] The coexpression of a *Hevea* truncated HMG-CoA reductase and a *Brassica* C24-sterol methyltransferase type 1, a critical enzyme in sterol biosynthesis, enhanced phytosterol content by 2.4-fold in transgenic tobacco seeds.[46]

Enfissi et al.[47] have achieved production of more nutritious tomato fruits that are enriched in health-promoting isoprenoids by the overexpression of an Arabidopsis HMG-CoA reductase of the MVA pathway, or that of a bacterial 1-deoxy-D-xylulose-5-phosphate synthase, an enzyme of the MEP pathway. Tomatoes that overexpressed Arabidopsis HMG-CoA reductase showed a 2.4-fold increase in phytosterols, while tomatoes that overexpressed plastid-targeted *Escherichia coli* DXP synthase exhibited a 1.6-fold increase in carotenoid content. This experiment illustrated that the manipulation of DXP synthase expression affected the plastid-localized MEP pathway in transgenic tomato.[47]

The cloning of a cDNA encoding a cytochrome P450, CYP71AV1, presents the initial step toward engineering plants and microbes for enhanced artemisinin production. CYP71AV1, a trichome-specific sesquiterpene monooxygenase, catalyzes the oxidation of amorpha-4, 11-diene, a precursor in artemisinin biosynthesis.[48] This anti-malarial drug is present only at low levels in *A. annua,* and harvesting artemisin from transgenic plants or microbes that overexpress CYP71AV1 and other relevant genes would be the strategy to implement toward increasing artemisinin supply.

8.9 CAN PLANTS BE GENETICALLY ALTERED TO PRODUCE ISOPRENOIDS THAT PROMOTE PLANT DEFENSE AGAINST HERBIVORES?

Caterpillar herbivory induces plants to release terpenoids and other volatiles, attracting parasitoids or predators of the insect herbivore and thereby forming an indirect mechanism of plant defense.[49] Hence, it is not surprising that insect salivary factors have been identified to suppress the expression of genes involved in isoprenoid biosynthesis, particularly those in the MEP pathway.[50] The genetic manipulation of terpenoid production in transgenic Arabidopsis has resulted in the attraction of insect bodyguards to the plant, the carnivorous mites *Phytoseiulus persimilis*.[51] Furthermore, the expression of a maize gene encoding the enzyme terpene synthase TPS10 that synthesizes herbivory-induced sesquiterpenes in transgenic *Arabidopsis* enhanced the attraction of females of the parasitoid *Cotesia marginiventris* to lepidopteran larvae.[52] These are just some examples in which the manipulation of isoprenoid biosynthesis leading to an overproduction of terpenoids has successfully attracted the predator of the insect herbivore, indirectly conferring protection against the herbivore.

8.10 WHAT OTHER ISOPRENOIDS ARE IMPORTANT IN INDUSTRY?

Natural rubber is made up of isoprenoids and there are many plants that produce natural rubber. For example, the chicle tree is the plant from which the brand name of the chewing gum "Chiclets" is derived. Although the use of natural rubber from plants has decreased over the past decades due to the production of synthetic rubber, much natural rubber is still being used in the production of tires, latex gloves and condoms.

Polyterpenes like natural rubber are generated from IPP by the action of the enzyme rubber transferase. These include *cis*-1,4-polyisoprene from the rubber tree *Hevea brasiliensis* and the desert shrub guayule (*Parthenium argentatum*) and *trans*-1, 4-polyisoprene (gutta-percha) from *Palaquium gutta*.[53,54] Gutta-percha is a good non-electrical conductor that is used for the insulation of electric cables and wires. Historically, it was utilized in the manufacture of the first trans-Atlantic cable and was used in the core of golf balls. Its disadvantage over *Hevea* natural rubber lies in its reduced elasticity due to its *trans* configuration.

Since *Hevea* latex contains a significant amount of *cis*-1,4-polyisoprene (30–50% w/w), several investigations have been carried out to understand the biosynthesis of *cis*-1,4-polyisoprene in its laticifers, the specialized ducts adjacent to the phloem of this tree.[54–56] Ko et al.[55] reported that an analysis of the laticifer transcriptome revealed not only an abundance of genes encoding rubber particle proteins but also genes related to defense. This is not unexpected given that *cis*-1,4-polyisoprene is synthesized on the surface of the rubber particles. An earlier study focusing on the gene expression in *Hevea* had shown an enrichment of transcripts related to plant defense and rubber biosynthesis in laticifers.[54]

Other commercially important isoprenoids are rosin and turpentine, both of which are derived from the distillation of conifer resin. Turpentine is the distillate, while rosin is the solid residue remaining following distillation. Rosin is utilized for the production of inks, soaps, adhesives, sealing wax and varnishes. Its use on string instruments improves their sound effects. It is also used by certain performers to increase friction on ballet shoes and on the ropes and gloves of bull riders. Turpentine, which largely consists of monoterpenes, has industrial applications as a solvent for thinning oil-based paints, in varnishes and in wood preservatives. It is also used as an ingredient in household cleaners because of its antiseptic properties.

8.11 CONCLUSIONS

Isoprenoids are a large, assorted and significant group of compounds that greatly affect our lives. This is largely due to their diversity in structure and function. Their abundance and diversity encountered in the plant kingdom alone is evident when we compare useful plant isoprenoids from species ranging from *Taxus* (taxol), conifers (turpentine, rosin) to *Hevea* (rubber). Despite conservation in each of the MVA and the MEP pathways in plants, the final reactions will yield the already mentioned diverse products as well as carotenoids, isoprene, taxolchlorophylls, gibberellins and sterols.

With the discovery of the second (MEP) pathway leading to IPP biosynthesis, scientists will address issues related to the regulation of these two independent pathways and as to how crosstalk occurs between them. Scientists will utilize techniques in reverse genetics, metabolic profiling and posttranscriptional gene silencing to this end. A deeper understanding of these pathways from investigations using these relatively new approaches will eventually lead to better strategies in the manipulation of these pathways to benefit mankind.

Investigations on relatively new terpenoids from marine organisms have demonstrated that some of them have unusual structures, e.g., frequent substitutions with halogen atoms, while others have demonstrated potential use as biomedical drugs.[57,58] Although much is already known on isoprenoids, there is still much more to be discovered in this expanding field.

REFERENCES

1. Sacchettini, J.C. and Poulter, C.D., Creating isoprenoid diversity, *Science*, 277, 1788–1789, 1997.
2. Eisenreich, W., Rohdich, F., and Bacher, A., Deoxyxylulose phosphate pathway to terpenoids, *Trends Plant Sci.,* 6, 78–84, 2001.
3. Penuelas, J. and Munne-Bosch, S., Isoprenoids: An evolutionary pool for photoprotection, *Trends Plant Sci.*, 10, 166–169, 2005.
4. Owen, S.M. and Penuelas, J., Opportunistic emissions of volatile isoprenoids, *Trends Plant Sci.*, 10, 420–426, 2005.
5. Cal, K., Skin penetration of terpenes from essential oils and topical vehicles, *Plant Med.*, 72, 311–316, 2006.
6. Edwards, P.A. and Ericsson J., Sterols and isoprenoids: Signaling molecules derived from the cholesterol biosynthetic pathway, *Annu. Rev. Biochem.*, 68, 157–185, 1999.
7. Maxfield, F.R. and Tabas, I., Role of cholesterol and lipid organization in disease, *Nature*, 438, 612–621, 2005.
8. Maxfield, F.R. and Menon, A.K., Intracellular sterol transport and distribution, *Curr. Op., Cell Biology*, 18, 379–385, 2006.
9. Chappell, J., Biochemistry and molecular biology of the isoprenoid biosynthesis pathway in plants, *Ann. Rev. Plant Physiol. Plant Mol. Biol.*, 46, 521–547, 1995.
10. Lichtenthaler, H.K. et al, Biosynthesis of isoprenoids in higher plant chloroplasts proceeds via a mevalonate-independent pathway, *FEBS Letts.*, 400, 271–274, 1997.
11. Rohmer, M., The discovery of a mevalonate-independent pathway for isoprenoid biosynthesis in bacteria, alga and higher plants, *Nat. Prod. Rep.*, 16, 565–574, 1999.
12. Seto, H., Watanabe, H., and Furihata, K., Simultaneous operation of the mevalonate and non-mevalonate pathways in the biosynthesis of isopentenyl diphosphate in *Streptomyces aeriouvifer, Tetrahedron Lett.*, 31, 7979–7982, 1996.
13. Kasahara, H. et al., Contribution of the mevalonate and methylerythritol phosphate pathways to the biosynthesis of gibberellins in Arabidopsis, *J. Biol. Chem.*, 277, 45188–45194, 2002.
14. Nagata, N. et al., Mevalonic acid partially restores chloroplast and etioplast development in *Arabidopsis* lacking the non-mevalonate pathway, *Planta*, 216, 345–350, 2002.

15. Laule, O. et al., Cross-talk between cytosolic and plastidial pathways of isoprenoid biosynthesis in *Arabidopsis thaliana*, *Proc. Natl. Acad. Sci. USA*, 100, 6866–6871, 2003.
16. Hemmerlin, A. et al., Cross-talk between the cytosolic mevalonate and the plastidial methylerythritol phosphate pathways in tobacco bright yellow-2 cells, *J. Biol. Chem.*, 278, 26666–26676, 2003.
17. Hemmerlin, A. and Bach, T.J., Effects of mevinolin on cell cycle progression and viability of tobacco BY-2 cells, *Plant J.*, 14, 65–74, 1998.
18. Hemmerlin, A., Brown, S.C., and Bach, T.J., Function of mevalonate in tobacco cell proliferation, *Acta Bot. Gall.*, 146, 85–100, 1999.
19. Bach, T.J. et al., Mevalonate biosynthesis in plants, *Crit. Rev. Biochem. Mol. Biol.*, 34, 107–122, 1999.
20. Seemann, M. et al., Isoprenoid biosynthesis in plant chloroplasts via the MEP pathway: Direct thylakoid/ferredoxin-dependent photoreduction of GcpE/IspG, *FEBS Letts.*, 580, 1547–1552, 2006.
21. Bertea, C.M. et al., Isoprenoid biosynthesis in *Artemisia annua*: Cloning and heterologous expression of a germacrene A synthase from glandular trichome cDNA library, *Arch. Biochem. Biophys.*, 448, 3–12, 2006.
22. Graβmann, J., Terpenoids as plant antioxidants, *Vitamins and Hormones* 72, 505–535, 2005.
23. Xiao, W.L. et al., Triterpenoids from *Schisandra lancifolia* with anti-HIV-1 activity, *J. Nat. Prod.*, 69, 277–279, 2006.
24. Liu, X.T., et al., *ent*–Rosane and labdane diterpenoids from *Saggittaria sagittifolia* and their anti-bacterial activity against three oral pathogens, *J. Nat. Prod.*, 69, 255–260, 2006.
25. Roy, A. and Saraf, S., Limonoids: Overview of significant bioactive triterpenes distributed in plant kingdom, *Biol. Pharm. Bull.*, 29, 191–201, 2006.
26. van Schie, C.C.N., Haring, M.A., and Schuurink, R.C., Regulation of terpenoid and benzenoid production in flowers, *Curr. Opin. Plant Biol.*, 9, 203–208, 2006.
27. Bouvier, F. et al., Oxidative tailoring of carotenoids: A prospect toward novel functions in plants, *Trends in Plant Sci.,* 10, 187–194, 2005.
28. Loreto, F. and Velikova, V., Isoprene produced by leaves protects the photosynthetic apparatus against ozone damage, quenches ozone products, and reduces lipid peroxidation of cellular membranes, *Plant Physiol.*, 127, 1781–1787, 2001.
29. Penuelas, J. and Llusia, J., Plant VOC emissions: Making use of the unavoidable, *Trends Ecol. Evol.*, 19, 402–404, 2004.
30. Sharkey, T.D. and Singsaas, E.L., Why plants emit isoprene, *Nature*, 374, 769, 1995.
31. Sharkey, T.D. and Yeh, S., Isoprene emission from plants, *Annu. Rev. Plant. Physiol. Plant Mol. Biol.*, 52, 407–436, 2001.
32. Ling, W.H. and Jones, P.J.H., Minireview dietary phytosterols: A review of metabolism, benefits and side effects, *Life Sci.,* 57, 195–206, 1995.
33. Tikkanen, M.J., Plant sterols and stanols, *Handb. Exp. Pharmacol.*, 170, 215–230, 2005.
34. Weststrate, J.A. and Meijer, G.W., Plant sterol-enriched margarines and reduction of plasma total- and LDL-cholesterol concentrations in normocholesterolaemic and mildly hypercholesterolaemic subjects, *Eur. J. Clin. Nutr.*, 52, 334–343, 1998.
35. Hussein, G. et al., Astaxanthin, a carotenoid with potential in human health and nutrition, *J. Nat. Prod.*, 69, 443–449, 2006.

36. Rodriguez-Rodriguez, R. et al., Triterpenic compounds from "orujo" olive oil elicit vasorelaxation in aorta from spontaneously hypertensive rats, *J. Agri. Food Chem.*, 54, 2096–2102, 2006.
37. Alberts, A.W. et al., Mevinolin: A highly potent competitive inhibitor of hydroxymethylglutaryl-coenzyme A reductase and a cholesterol-lowering agent, *Proc. Natl. Acad. Sci. USA*, 77, 3957–3961, 1980.
38. Shigi, Y., Inhibition of bacterial isoprenoid synthesis by fosmidomycin, a phosphonic acid-containing antibiotic, *J. Antimicrob. Chemother.*, 24, 131–145, 1989.
39. Jomaa, H. et al., Inhibitors of the nonmevalonate pathway of isoprenoid biosynthesis as antimalarial drugs, *Science*, 285, 1573–1576, 1999.
40. Kuntz, L. et al., Isoprenoid biosynthesis as a target for antibacterial and antiparasitic drugs: Phosphonohydroxamic acids as inhibitors of deoxyxylulose phosphate reducto-isomerase, *Biochem. J.*, 386, 127–135, 2005.
41. Swanson, K.M. and Hohl, R.J., Anti-cancer therapy: Targeting the mevalonate pathway, *Curr. Cancer Drug Targets*, 6, 15–37, 2006.
42. Chye, M.L., Tan, C.T., and Chua, N.H., Three genes encode 3-hydroxy-3-methylglutaryl coenzyme A reductase in *Hevea brasiliensis*: *hmgl* and *hmg3* are differentially expressed, *Plant Mol. Biol.*, 19, 473–484, 1992.
43. Schaller, H. et al., Expression of the *Hevea brasiliensis* 3-hydroxy-3-methylglutaryl coenzyme A reductase 1 in *Nicotiana tabacum* results in sterol overproduction, *Plant Physiol.*, 109, 761–770, 1995.
44. Harker, M. et al., Enhancement of seed phytosterol levels by expression of an N-terminal truncated *Hevea brasiliensis* (rubber tree) 3-hydroxy-3-methylglutaryl-CoA reductase, *Plant Biotech. J.*, 1, 113–121, 2003.
45. Chappell, J. et al., Is the reaction catalyzed by 3-hydroxy-3-methylglutaryl coenzyme A reductase a rate limiting step for isoprenoid biosynthesis in plants? *Plant Physiol.*, 109, 1337–1343, 1995.
46. Holmberg, N. et al., Co-expression of N-terminal truncated 3-hydroxy-3-methylglutaryl CoA reductase and C24-sterol methyltransferase type 1 in transgenic tobacco enhances carbon flux toward end-product sterols, *Plant J.*, 36, 12–20, 2003.
47. Enfissi, E.M.A. et al., Metabolic engineering of the mevalonate and non-mevalonate isopentenyl diphosphate-forming pathways for the production of health-promoting isoprenoids in tomato, *Plant Biotech. J.*, 3, 12–27, 2005.
48. Teoh, K.H. et al., *Artemisia annua* L. (Asteraceae) trichome-specific cDNAs reveal CYP71AV1, a cytochrome P450 with a key role in the biosynthesis of the antimalarial sesquiterpene lactone artemisinin, *FEBS Letts.*, 580, 1411–1416, 2006.
49. Kessler, A. and Baldwin, I.T., Plant responses to insect herbivory: The emerging molecular analysis, *Annu. Rev. Plant Biol.*, 53, 299–328, 2002.
50. Bede, J.C. et al., Caterpillar herbivory and salivary enzymes decrease transcript levels of *Medicago truncatula* genes encoding early enzymes in terpenoid biosynthesis, *Plant Mol. Biol*, 60, 519–531, 2006.
51. Kappers, I.F. et al., Genetic engineering of terpenoid metabolism attracts bodyguards to *Arabidopsis*, *Science*, 309, 2070–2072, 2005.
52. Schnee, C. et al., The products of a single maize sesquiterpene synthase form a volatile defense signal that attracts natural enemies of maize herbivores, *Proc. Natl. Acad. Sci. USA*, 103, 1129–1134, 2006.
53. Cornish, K., Similarities and differences in rubber biochemistry among plant species, *Phytochemistry*, 57, 1123–1134, 2001.
54. Kush, A. et al., Laticifer-specific gene expression in *Hevea brasiliensis* (rubber tree), *Proc. Natl. Acad. Sci. USA,* 87, 1787–1790, 1990.

55. Ko, J.H., Chow, K.S., and Han, K.H., Transcriptome analysis reveals novel features of the molecular events occurring in the laticifers of *Hevea brasiliensis* (para rubber tree), *Plant Mol. Biol.*, 53, 479–492, 2003.
56. Mau, C.J.D. et al., Protein farnesyltransferase inhibitors interfere with farnesyl disphosphate binding by rubber transferase, *Eur. J. Biochem.*, 270, 3939–3945, 2003.
57. Newman, D.J. and Craig, G.M., Marine natural products and related compounds in clinical and advanced preclinical trials, *J. Nat. Prod.*, 67, 1216–1238, 2004.
58. Konig, G.M., Natural products from marine organisms and their associated microbes, *ChemBioChem*, 7, 229–238, 2005.

Section II

Demonstrations and Experiments in Lipid Science

9 WHAT'S IN A POTATO CHIP?

Robert G. Ackman, Anne Timmins and Suzanne M. Budge*

CONTENTS

9.1 RATIONALE

The media have discovered obesity in Canadian children. This stems from a report, "Healthy Weights for Healthy Kids," tabled by a House of Commons committee in late March of 2007. The incidence of obesity in children of the 2–17-years age group has jumped to 26%. Inactivity can be implicated, but modern snack foods and drinks are really major factors. The sugar contents of the latter are invisible and, in many packages, the content information is in small print and often printed sideways to make sure nobody reads it.

Potato chips, a popular food, are easily seen and handled. Each bag has a prominent table of "Nutrition Facts" that is usually ignored irrespective of whether the purchaser is alone or in one of the groups where young people gather socially. As shown in 9.1, a serving is (60 g) and fat is 22 g of that (33%). Can a youths visualize 60 g or 22 g? This is doubtful, as is their grasp of the identity of the fat, sunflower oil in this case. The fact that it is a fluid fat allows us to recover and visually present that oil. A small group approach fits exactly the popular social behavior of the teenagers, and the whole operation as a bonus has a scientific twist illustrating densities of different fluids.

9.2 A VISUAL DEMONSTRATION OF FAT EXTRACTION AND FLUID DENSITIES

There is a need to show students just how much fat they are eating in popular snack foods. This demonstration was developed for high school classes touring a university

*Corresponding author

Figure

Nutrition Facts **Valeur nutritive** Per 1 package (60g) pour 1 paquet (60g)	
Amount Teneur	% Daily Value % valeur quotidienne
Calories / Calories 330	
Fat / Lipides 22 g	33%
Saturated / saturés 2 g + Trans / trans 0.2 g	12%
Cholesterol / Cholestérol 0 mg	0%
Sodium / Sodium 400 mg	17%
Carbohydrate / Glucides 31 g	10%
Fibre / Fibres 1 g	5%
Sugars / Sucres 0 g	
Protein / Protéines 4 g	
Vitamin A / Vitamine A	0%
Vitamin C / Vitamine C	25%
Calcium / Calcium	2%
Iron / Fer	4%

INGREDIENTS: SPECIALLY SELECTED POTATOES, SUNFLOWER OIL, SALT.

INGRÉDIENTS: POMMES DE TERRE SPÉCIALEMENT SÉLECTIONNÉES, HUILE DE TOURNESOL, SEL.

FIGURE 9.1 Typical nutrition information to be found on a bag of potato chips in Canada

food science department. It is based on the common potato chip because this snack food is popular, cheap and readily available. As a bonus, no matter how large a bag is used, there is seldom any residue to discard.

A major visual benefit is that the results are conveniently measured by volume in glass, and the behaviors of fluids of different densities can be observed. Additional benefits are achieved by forming teams of up to three or four students. Depending on numbers, each person in a team can do one step in the procedure, or competition interest can be achieved with different teams' having different products. In any case, spectators should be kept at a distance of 3 to 5 feet (1–2 meters). The initial steps are rapid, the filtration slow and the series of water additions also slow, if held long enough to be effective. The amount of glassware needed is not extensive. Plastics should be avoided. There is a remote risk of fire and of inhaling acetone vapor. If enough fume hood space is not available, a spacious well-ventilated room may be safe if there are no electrical sparks. Acetone vapor is inflammable but relatively harmless. However, although it evaporates rapidly, spills should be wiped up with paper toweling. Extracted chips and filter paper should be disposed of outdoors.

9.2.1 Procedure for Fat Separation and Recovery

9.2.1.1 Introduction

This is a simple lab designed to show students the amount of fat that they consume from a bargain bag of chips (which is about three servings or 150 g). The fat is extracted with acetone, and water is added to facilitate the phase separation.

9.2.1.2 Materials

- Chips
- Glass mortar and pestle
- Acetone
- Water
- Balance
- Beaker (400 ml)
- Stirring rod
- Graduated cylinder (250 ml), preferably with glass stopper
- Funnel stand
- Funnel and fluted filter paper

9.2.1.3 Procedure

1. Weigh 50 g of chips and place in mortar. Grind with the pestle.
2. Transfer to the beaker. Add 50 ml acetone. Rinse the mortar and pestle with another 50 ml acetone and add that to the beaker. Stir for 3 minutes.
3. Filter chips and acetone mixture into the graduated cylinder. Add water to the cylinder in 10 ml aliquots as instructed to facilitate stepwise phase separation.
4. After each addition of water, shake gently and let stand for a few minutes. Approximately 50 ml of water is eventually required for complete phase separation.
5. After separation is complete, record the volume of fat.

9.2.1.4 Results and Discussion

1. The density of water can be taken as 1.000, that of acetone as 0.790 and of canola oil as 0.927. Explain the observations on each stage of water addition.
2. Assume the recovery of fat from this simple extraction is 80%. How does the mass recovered compare with the nutrient facts on the label?
3. How many fire extinguishers were there in the room? (A marginal question to find out how observant students are.)
4. Where can one find acetone in the ordinary household (nail polish remover)?

10 Oils and Fats
Simple Demonstrations of Properties and Uses

Thomas A. McKeon

CONTENTS

10.1 INTRODUCTION

In 1993 Drs. Ladell Crawford, Katrina Cornish and Glenn Fuller initiated the Agricultural Sciences Academic Workshop at the Western Regional Research Center. The purpose of the workshop is to introduce third-year high school students to a broad variety of scientific experience, both as a community outreach effort and in the hope of encouraging some to become interested in scientific careers. The two local schools from which students are recruited include one with a high percentage of minority students traditionally underrepresented in the sciences. The material included in this chapter is presented as one 90-minute session of the approximately 15 sessions that the students attend. Additional information on this workshop can be found at <http://www.ars.usda.gov/is/pr/2005/050209.o.albany.htm>.

10.2 FAT IN THE DIET

Most of the students are well-informed and have heard news reports describing the health effects of certain fatty acids. The intention of the following discussion is to illustrate some of the points that could help them understand what the news reports are describing and how the different fatty acids play a role in maintaining health.

There is a considerable amount of information about dietary fat in the popular media. Much of this information is related to health effects of fats, but the meaning of terms presented in such articles is usually lacking. Yet, it is precisely the meaning of such terms that helps an informed consumer make appropriate dietary decisions. Examples of these terms include "*trans*-fats," "saturated fats" and "omega-3 fats." The structures of fatty acids that are frequently described in press releases are presented in 10.1. Although the student participants have all taken chemistry

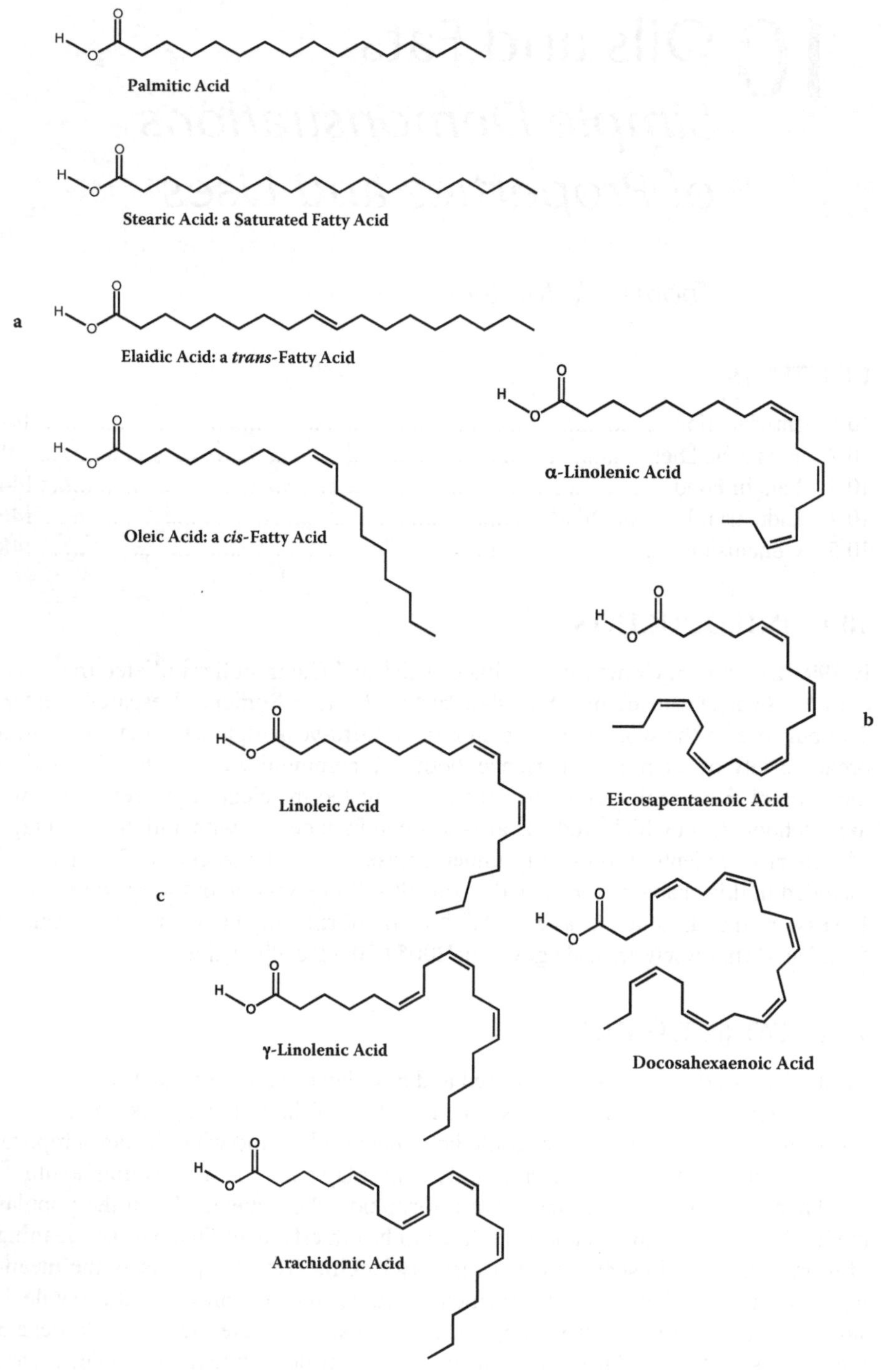

Palmitic Acid

Stearic Acid: a Saturated Fatty Acid

a

Elaidic Acid: a *trans*-Fatty Acid

Oleic Acid: a *cis*-Fatty Acid

α-Linolenic Acid

Eicosapentaenoic Acid

b

Linoleic Acid

c

Docosahexaenoic Acid

γ-Linolenic Acid

Arachidonic Acid

FIGURE 10.1 Common dietary fatty acids. (a) Saturated and monounsaturated (*cis* and *trans*), (b) omega-3 polyunsaturated and (c) omega-6 polyunsaturated.

courses, they often have very limited exposure to organic structures, especially for molecules with as many atoms as a typical fatty acid. Graphical depictions of fatty acid structures are often difficult to visualize, so it is important to have something concrete that the students can handle. One choice is wire models; space-filling models are also useful. The main points to be illustrated are the similar structures of saturated and *trans* fatty acids, the fact that fatty acids with *cis* double bonds are the common unsaturated fatty acids and the effect of increasing numbers of *cis* double bonds on the structure of unsaturated fatty acids. It should also be pointed out to the students that molecules are not rigid; they generally rotate freely about single bonds, so that the difference between saturated and unsaturated fatty acids is some restricted rotation as limited by the rigid double bond.

There continues to be some controversy with regard to whether *trans* fats have a similar effect on health as saturated fats, or if they are detrimental to health. The purpose of the lecture is not to proselytize, but to present information that will give the students the background to make informed decisions. Because the effects of *trans* fatty acids remain controversial, this situation is generally described as follows:

> The USDA dietary recommendation for saturated fatty acids in the diet is to consume less than 10% of the caloric intake, while total fat intake is recommended at a level of 20–35%, depending on age and other factors http://www.health.gov/dietaryguidelines/dga2003/recommendations.htm. Saturated fats are found in dairy and meat products at a high percent of fat content, and are present in plant fats (vegetable oils) generally at lower levels. Among the commonly consumed plant fats and oils, however, palm oil, coconut oil and cocoa butter (the fat present in chocolate) are high in saturated fat content.

The *trans* fatty acids occur naturally in dairy products and in meat, to a very small extent. They also occur naturally in some non-food seed oils, such as tung oil, used as a drying oil, i.e., an oil that gradually oxidizes in air to provide an impenetrable coating on materials such as wood. The major sources of *trans* fatty acids in the diet are oils that have been partially hydrogenated to produce a solid fat. Cottonseed, corn and soy oils are most often used to produce such vegetable shortenings, and lard, which is solid fat from animals, may also be partially hydrogenated to improve stability. Due to the geometry of the *trans* double bond, the structure and physical properties of *trans* fatty acids resemble those of saturated fatty acids while, chemically, they are unsaturated fatty acids. The *trans* fatty acid elaidic acid has a melting point of 43°C and forms a solid fat at room temperature, as does stearic acid with a melting point of 68°C, while the *cis* fatty acid oleic acid has a melting point of 13°C and forms a (liquid) oil at room temperature. Thus, prior to the requirement for specifically labeling the *trans* fat content of foods, they would be included with the unsaturated fat content, and one would have to assume their presence if the food contained a partially hydrogenated fat. The perceived benefit of these fats was as an oxidatively stable replacement in baked or fried products for lard or butter, which contain cholesterol.

The beneficial unsaturated fatty acids contain *cis* double bonds that dramatically alter the structure when compared with saturated fatty acids. The melting points of a fat will be lower with increasing proportion of unsaturated to saturated fatty acids or higher content of polyunsaturated fatty acids. While olive oil is often considered to be one of the better food oils for health, other vegetable oils contain greater amounts of the essential fatty acids linoleic (omega-6) and linolenic (omega-3) acids. In recent years, the fatty acids eicosapentaenoic acid (EPA) and docosahexaenoic acid (DHA), both omega-3, have been shown to be especially important in fetal brain development and maintenance of healthy neural and vascular systems. The dietary source of these fatty acids is fish, though they are actually derived from the algae and phytoplankton that fish feed on. Arachidonic acid (AA), an omega-6, is important in balance with EPA and DHA, also playing a role in neural and vascular responses. Dietary sources of AA include some animal products and some fungi. The DHA and EPA are also produced in humans from alpha-linolenic acid, but not efficiently, while AA is produced from gamma linolenic acid, which is present in oils from borage seed, evening primrose seed and blackberry seed.

Fat consumption continues to be a focus of health research, with excessive consumption of saturated and *trans* fats associated with type 2 diabetes, metabolic syndrome, fatty liver, elevated cholesterol associated with low density lipoprotein (LDL cholesterol), reduced cholesterol associated with high density lipoprotein (HDL cholesterol), and atherosclerosis. On the other hand, consumption of oils containing beneficial fatty acids is associated with improved neural and circulatory condition, reduced levels of chronic inflammation, improved skin health and better health overall. It is important to point out that the results of research on health benefits of different fatty acids frequently receive publicity and continue to progress, and occasionally the conclusions drawn from this research contradict previous conclusions.

10.3 FATS IN FOOD

Fat adds flavor, texture and calories to food. It also provides a cooking medium for frying, sautéeing and deep-fat frying. This latter medium can add considerably to the fat content of the product and, depending on the oil used, provide beneficial or detrimental nutritional quality. The fat content of a food product can be simply illustrated by extracting the fat from potato chips. The fat extraction can be done in different ways, depending on the time devoted to it. We have used solvent extraction and supercritical CO_2 extraction of fat. The latter apparatus is not likely to be available in many laboratories, so the extraction procedure described is the most general approach. Acetone (all laboratory experiments are conducted with students and instructor wearing laboratory coats, protective eye wear, and gloves when needed) is used as the extraction solvent and the extraction can take two forms:

1. Individual experiments, where students are each given a tube with 1 g of crushed chip. They add 5 ml of acetone, extract in a capped tube heated for 30 min, then centrifuge, remove the solvent and evaporate it in a hood under a stream of nitrogen. We do this qualitatively, and a high-fat content chip leaves a significant oily residue. This can be done quantitatively by letting the students weigh the chip and then the extracted oil. The comparisons that can be done include deep-fat-fried chips and baked (low-fat), or chips fried in partially hydrogenated fats.

2. For a more graphic display, a larger portion of chips, 25 g, is extracted in 250 ml acetone under reflux, the extract filtered and solvent removed by rotary evaporation. This approach has the advantage of generating a considerable volume of oily residue and allows a good visual comparison of the difference between deep-fried and baked chips.

The reader is referred to Chapter 9 for a detailed protocol on the extraction of fat from potato chips.

Some lipids have properties that allow mixing of normally immiscible liquids. This property is called emulsification, and one naturally occurring emulsifier is phospholipid, a major component of egg yolks, as well as a minor component of seed oils prior to refining. The common name for phospholipid is lecithin. One example of the use of egg yolk as an emulsifier is in making mayonnaise. The most convenient way to make mayonnaise in a laboratory setting is by blender, adding lemon juice or vinegar to the blender, egg yolk and a small amount of cream of tartar, and slowly adding oil until it thickens to the proper consistency. The volume of oil needed depends on the volume of egg yolk (lecithin content), vinegar or lemon juice, the acidity and possibly the type of oil used. As a rough guide, the egg yolk from a large egg will emulsify approximately 200 ml oil. Because of the potential hazard of *Salmonella* contamination of the egg, it is not a good idea to sample the mayonnaise. If it is desired to sample the mayonnaise, the egg yolks should be pasteurized by combining them with water, vinegar, and lemon juice and heating briefly to 65°C as described on http://www.chefdecuisine.com/course/sauce/MAYONNAISE.asp. Alternatively, pasteurized egg yolks can be purchased.

Fat also affects the texture of many foods, and one of the most desired textures is the "melt in the mouth" feeling. One simple, and always very popular, example of this phenomenon is chocolate. The chocolate used in this experiment is limited to that containing only cocoa butter, not dairy butter, vegetable oils or partially hydrogenated fats. This is mainly to limit the discussion to the unusual composition of cocoa butter, which is primarily composed of a single species of triacylglycerol containing palmitate, oleate and stearate, with oleate mainly in the *sn*-2 position (Figure 10.2). As a result of this arrangement, there is high saturated fat content and the melting point of cocoa butter is relatively high, corresponding to near body temperature. Moreover, the fact that a single species of triacylglycerol makes up most of the cocoa butter gives it a relatively sharp melting point, which disperses the flavor components as the chocolate melts in the mouth. The combination of mouth feel and flavor account for the popularity of chocolate. While recent reports ascribe

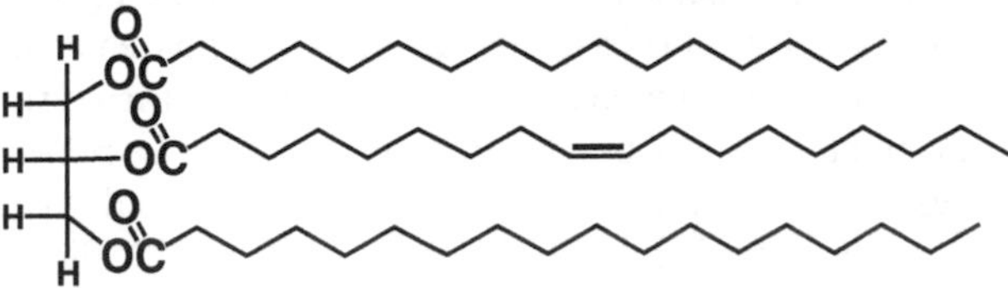

FIGURE 10.2 Major component of cocoa butter:1-palmitoyl-, 2-oleoyl-, 3-stearoyl-glycerol.

health benefits to dark chocolate, the high saturated fatty acid content must also be kept in mind.

Although the following experiment has not been included in the curriculum, it is one of the demonstrations put on by the local American Chemical Society for middle school students and parents during National Chemistry Week. It also illustrates the value of fat as a delivery system for flavor. The recipe for liquid nitrogen ice cream was provided by Dr. Alex Madonik of the American Chemists Society, California Section, National Chemistry Week organizer. There are also similar recipes available on the Internet.

The following recipe produces 1.25 gallons of ice cream. Because liquid nitrogen is used to freeze the mix, protective gloves and eyewear must be worn.

4 cups of sugar
1 quart half and half
1 quart whole milk
1 pint whipping cream
2 tablespoons lemon juice
½ cup frozen orange juice concentrate
2 10-ounce packages of frozen raspberries

The ingredients are combined, then 4–5 liters of liquid nitrogen are added with rapid stirring. The rapid freezing prevents formation of large ice crystals and provides a finely dispersed solid suspension of fat particles and aqueous flavor components that impart the creamy sensation with flavor delivery.

10.4 INDUSTRIAL USES OF OILS

Probably the major nonfood use of fats and oils is in production of soaps and detergents. Students in this course especially value hands-on experiments, and the soap produced is used by the students to wash their hands. As ordinary as this may seem, seeing soap made and "testing" the end result is one of the more popular experiments. It can also be pointed out that soaps form a deposit in hard water and are not as effective, while many detergents can foam effectively in the presence of the minerals that cause water hardness.

To produce soap, we use coconut oil, because coconut soap produces plentiful suds, even in the presence of some salt, and has a mild, pleasant odor. Coconut oil contains about 50% lauric acid, a saturated fatty acid, and as a result is solid, so it must first be melted by warming its container in warm water. Then, 225 g of the liquid oil are added to a beaker, with 750 ml water and 48 g sodium hydroxide (NaOH). The amount of NaOH is based on the amount of base needed to hydrolyze (saponify) the coconut oil to completion, assuming it is all trilaurin (Figure 10.3). Because palmitic and oleic acids are also present, it should be remembered that the NaOH is in excess. The mixture is heated at 80°C and stirred rapidly on a heater equipped with magnetic stirring capability until the saponification is complete, at which point there will be no oil on the surface when stirring is stopped. This process takes 2–4 hours, depending on the rate of stirring. The soap is salted out by addi-

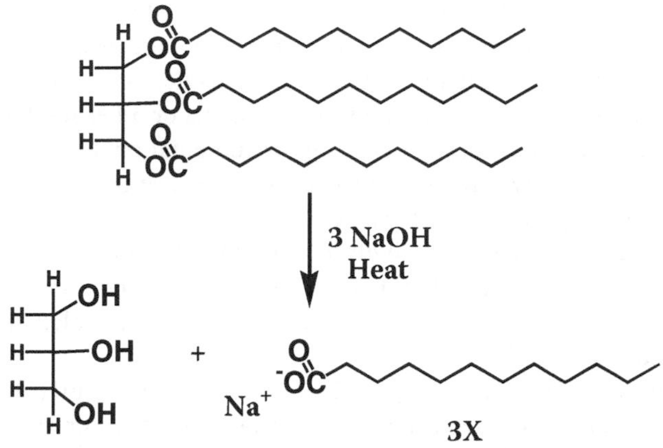

FIGURE 10.3 Saponification of trilaurin, a major component of coconut oil.

tion of 78 g sodium chloride (NaCl) with rapid stirring, then slow addition of 106 g NaCl. This is approximately a saturating amount of NaCl. The mixture is allowed to settle overnight, and the soap forms a solid layer on top of the salt solution. The soap is re-dissolved in 750 ml water by heating to 80°C and stirring on a heating magnetic stirrer, and salting out the soap as before. The soap block formed can be cut into smaller pieces that the students can use to wash their hands. For purposes of demonstration, the soap is prepared ahead of time, and on the day of the class, one reaction is brought to completion to demonstrate the salting-out process, and a second reaction is set up and heated just before the class begins, in order to show the early stage with a layer of oil on top of the container.

Soaps and detergents work because they interact with water and organic phases, thus helping to disperse organic soluble residues in water. Because water itself has a high surface tension, it does not readily interact with all types of stains, but soaps and detergents reduce the surface tension of water so that it can more readily interact with stains. In addition to washing hands with soap, the effect on surface tension can be illustrated by blowing bubbles. Bubble solutions are commercially available, but can also be made by adding equal parts of liquid dishwashing detergent, glycerol and water.

Another important industrial use of fats and oils is in lubricants. An important quality of lubricants is viscosity, and among the lubricants obtained from renewable resources, the viscosity of castor oil, as a result of its hydroxy fatty acid content, is very high compared with "normal" vegetable oils. Castor oil is 90% ricinoleate, 12-hydroxy oleic acid (Figure 10.4), and the interchain hydrogen bonding results in considerable "resistance to slipperiness," as Newton defined viscosity.

Falling ball viscometry is the simplest way of comparing the viscosity of castor oil with, e.g., canola oil. A toy marble, 1 cm in diameter, is of suitable density and

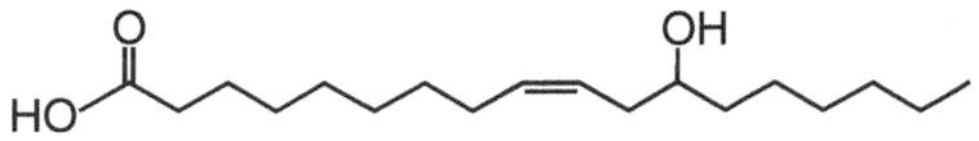

FIGURE 10.4 Ricinoleic acid.

size to use in the demonstration. 100- to 250-ml graduate cylinders are filled with castor or canola oil, marbles of the same size and weight are dropped simultaneously into each cylinder and the rate of fall observed. This demonstration is done strictly on a qualitative basis, as the rate of fall through canola or other food oils is too rapid to measure accurately without sophisticated instrumentation. On the other hand, the marble floats very slowly through the castor oil. Because both oils are safe, this is another hands-on experiment for the students.

Biofuels are becoming increasingly important worldwide. As petroleum fuels become more expensive and pollution a greater problem, fuels such as ethanol and biodiesel, which are derived from renewable resources, are supplementing gasoline, diesel fuel and heating oil.

Biodiesel can be made from vegetable oils by a process similar to soapmaking. We illustrate the ability to use vegetable oils or fats for fuel more simply by burning them shaped as a candle. Vegetable oils can be placed in a small cup or shotglass, a wick made from cotton string dipped in the oil, and the wick supported by pulling through aluminum foil covering the glass. A candle can be made from solid fat by inserting a wick into a container holding the fat. Any oil can be used in the glass; an aroma component can be added as well. We have made a candle from Crisco, which burns with a very smoky flame.

10.5 CONCLUSION

This course is designed to introduce students to a variety of scientific concepts that could spark their interest in science, or at least make science more relevant to their daily lives. The role of fats and oils in the diet as well as their nonfood uses can be illustrated through a number of simple, generally safe experiments that can be carried out at minimum expense.

11 Teaching Relationships between the Composition of Lipids and Nutritional Quality

Roman Przybylski

CONTENTS

11.1 INTRODUCTION

Lipids are the most difficult food components to assess for quality, and conversely, they affect the quality and nutritional value of foods the most. Lipid components are prone to oxidative degradation, generally leading to the formation of off-flavors and potential toxic compounds. Formed oxidation products may have detrimental health effects and stimulate free radical formation in living organisms. Yet nutritionally, many lipids are essential and required nutrients in our daily diet.

Teaching about lipid quality, and indirectly about food product quality and nutritional value, provides an opportunity to combine analytical data with the description of quality and nutritional factors. Standard analytical procedures offer a variety of assessments to evaluate and describe quality, level of degradation and nutritional value. Measuring the degree of oxidative degradation, including off-flavor components and composition of food lipids, provides data describing quality, safety and nutritional value of fats and oils, and, indirectly, foods containing these lipids. Chemical assessment of oxidative degradation such as peroxide value and *p*-anisidine value offers a partial quantification of primary and secondary oxidation products, respectively. However, the products measured by these methods are intermediates that further degrade to form a variety of products that are not assessed by these methods. Measuring off-flavor components provides data about the degree of oxidative degradation of oil or fat and is directly related to a sensory assessment ultimately used by consumers to assess oil quality.

The assessment of the nutritional factors of oils and fats is achieved by determining the amounts and composition of fatty acids, tocopherols, off-flavor components, and primary and secondary oxidation products. Quantifying the amount and composition of antioxidants offers information describing the extent of oxidative degradation, because these components are affected first by this process.

The exercises described here provide students with unique opportunities to learn how to perform analyses using standardized procedures, how to interpret data and how to find the degree of usefulness of analytical results in describing quality, nutritional value and potential health impact of lipids.

11.2 MATERIALS AND METHODS

11.2.1 Materials

The numbers of samples analyzed in these exercises are different for undergraduate and graduate students. Undergraduate students receive two coded fresh and aged oils for analysis; class size dictates whether students work individually or in groups. Graduate students always work individually and each student receives two sets of four coded fresh and aged oil and fat samples. All samples are assigned randomly to students to assure that each sample is run in quadruplication. Aged samples are prepared by storing fresh oils and fats under controlled conditions at 65°C for 5 days in open glass containers with surface-to-volume ratio at 1.

11.2.2 Methods

The goals for all these exercises is to make students aware that only standardized and approved methods are acceptable for quality assessment. In these assessments the following methods are utilized:

11.2.2.1 Fatty Acid Composition

In AOCS method Ce 1h-05[1] individual fatty acids and their isomers are assessed as methyl ester and the separations are carried on a gas chromatograph (GS) equipped with capillary column. Methyl heptadecenoate (17:1) or methyl heneicosanoate (21:0) are used as internal standards to quantify the fatty acid composition. The methylation procedure is modified to use sodium methoxide as catalyst. Individual fatty acid isomers are identified by comparing the retention time and relative retention data of separated peaks with that of standards using the same chromatographic conditions. For each set of analyzed samples, response factor for each fatty acid is established by running quantitative mixtures of standards at different concentration (AOCS Ce 1h-05).

11.2.2.2 Tocopherols

AOCS method Ce 8-89[2] was modified applying a different mobile phase containing 5% *tert* butyl methyl ether in hexane. This mobile phase provides very stable retention times and offers better separation of all tocopherol and tocotrienol isomers, including plastochromanol-8. The latter is a derivative of gamma tocotrienol with its side chain twice as long. Separation of the chromanols is performed on a normal phase HPLC using a column packed with silica (250 mm × 3.0 mm; 3 μm). A fluorescent detector is used with an excitation lamp operating continuously for detection of tocopherols. Fluorescent detectors utilizing strobe-type lamps for excitation requires a large sample of oils because the detector is 20 to 50 times less sensitive. For quantification, each individual tocopherol isomer needs to be calibrated separately due to its different response factor.

11.2.2.3 Peroxide Value

AOCS method Cd 8b-90[2] was used. In the standard method, during titration starch was used as the indicator, which was replaced by the potentiometric end-point measurement utilizing a platinum electrode. The end point is achieved when a reading of 75 mV is accomplished during titration.

11.2.2.4 *p*-Anisidine Value

AOCS method Cd 18-90[2] was used without any modification.

11.2.2.5 Flavor Components

Flavor components were analyzed using an internal trap system developed in our laboratory and installed in the GC.[3] In this system flavor components are transferred directly from oil samples onto SPB-5 capillary column (60 m × 0.32 mm, 0.5 μm phase thickness, Supelco), on which compounds are separated. The method eliminates intermediate steps and adsorbents usually used to isolate volatile components. For quantification, an internal standard such as dodecane in oil is applied. For all analyzed components the response factors are calculated similarly to that of fatty acids; see discussion above.

11.2.2.6 Sensory Analysis

Every student is trained to become a panelist, and attendance for the sensory sessions is obligatory. Oils are presented to sensory panelists in random order to assess the intensity of rancid odor. For evaluation of rancid odor intensity, a 15-point scale is used, where 0 represents no odor and 15 a very strong odor. For each assessment of rancid odor intensity a reference sample is used to provide reference intensity on scale. As reference, canola oil stored at 65°C for 6 days at surface-to-volume ratio of 0.5 is used to provide intensity of rancid odor at value of 8 on the scale used.

Graduate students additionally performed descriptive sensory assessment, using typical terminology usually applied for vegetable oils.

11.2.2.7 Students' Assignment

Undergraduate students analyze oils for fatty acid composition, peroxide value, flavor components composition and amount, and sensory assessment. Graduate students analyze oils for all parameters discussed for undergraduate students, plus they carry out identification of the oil or fat source using the composition of fatty acids and tocopherols as indicators. Also, these students perform statistical analysis based on fatty acid composition data to assess reproducibility and accuracy of the methods used.

TABLE 11.1
Fatty Acid Composition of Selected Oils

	Oil			
Fatty Acid (FA)	**Palm Olein**	**Canola**	**Mid Oleic Sunflower**	**Hydrogenated Canola**
C12:0	0.30			
C14:0	1.03			
C15:0	0.04			
C16:0	39.23	4.18	4.30	4.65
C16:1	0.19	0.23	0.09	0.17
C17:0	0.10	0.14	0.05	0.09
C18:0	4.15	1.85	3.71	4.23
C18:1*trans*	0.10	0.13	0.38	26.68
C18:1	42.73	60.64	59.56	51.89
C18:2*trans*	0.28	0.37	0.30	7.87
C18:2n-6	10.89	19.58	29.14	2.64
C18:3*trans*	0.07	1.60	0.07	0.40
C18:3n-3	0.20	8.49	0.68	0.14
C20:0	0.35	0.63	0.30	0.63
C20:1	0.14	1.30	0.26	1.08
C22:0	0.06	0.35	0.79	0.35
C24:0	0.07	0.15	0.26	0.16
C24:1	0.04	0.20	0.04	0.17
Groups				
trans	0.45	2.10	0.75	34.95
SAT	45.32	7.29	9.42	10.13
MUFA	43.11	62.36	59.96	53.31
PUFA	11.08	28.07	29.82	2.78
n-3	0.20	8.49	0.68	0.14
n-6	10.89	19.58	29.14	2.64
n-6/n-3	55	3	43	19

11.3 WORKING WITH RESULTS

Results from gas chromatography and HPLC analyses were calculated and tabulated to provide quantitative data for fatty acid composition, content of off-flavor components and tocopherols. Saturated, monounsaturated and polyunsaturated fatty acids (PUFAs) present in foods affect human metabolism differently, and also some fatty acids are essential. Omega-3 and omega-6 fatty acids share the same enzymes in their metabolism to eicosanoids and prostaglandins; the ratio between these two groups of essential fatty acids is an important factor to be considered in nutritional assessment.[4] Examples of some oils' composition are compiled in Table 11.1.

Oxidation, where mainly unsaturated fatty acids are involved and a variety of degradation products formed, is the main degradation process of lipids. Off-flavor components formed during oxidation are responsible for the rancid odor of oil that affects perception of quality negatively. The presence of off-flavor components is identified using gas chromatography by comparing the retention time and relative retention time to known standards, while their content using internal standards is determined. The actual chromatograms of the off-flavor components are usually very complex, and to make the interpretation easier, only the main off-flavor components are used for assessment as represented in Figure 11.1. These include hexanal, nonanal, total amount of off-flavor components and anisidine value; all represent the formation of secondary oxidation products in fresh and aged oils (Figure 11.1). Anisidine value is measuring nonvolatile carbonyl components formed as secondary oxidative degradation products; carbonyls are the main compounds of oxidative degradation. The same oils are assessed for intensity of rancidity by a sensory panel, and the data acquired are presented as a bar graph with their errors of estimation (Figure 11.2). Sensory data were statistically analyzed to assess the accuracy and

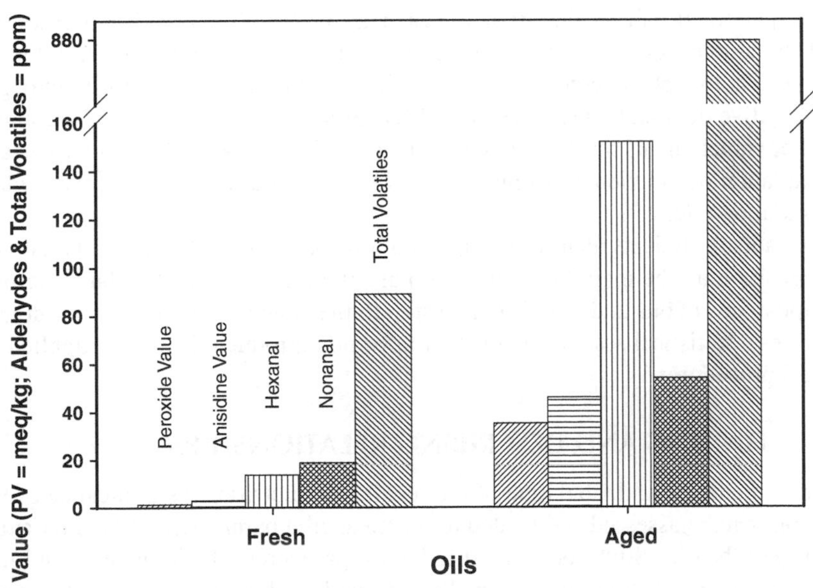

FIGURE 11.1 Formation of oxidative degradation products during canola oil aging.

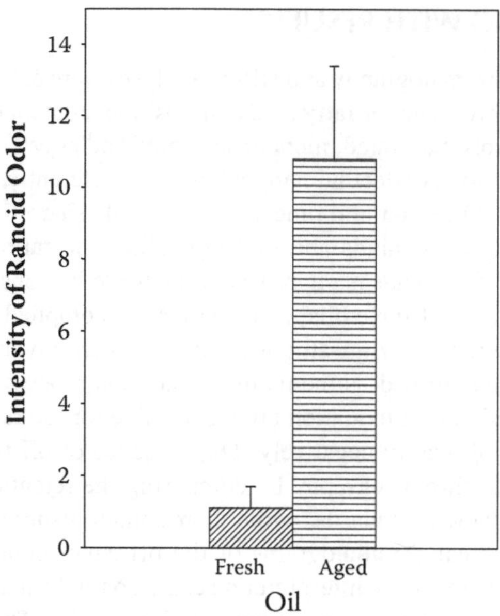

FIGURE 11.2 Intensity of rancid odor in canola oils. Error bars represent standard deviations for 4–12 replicates.

distribution. Graduate students performed descriptive sensory analyses to evaluate not only rancid odor intensity but also the type of other odors in the oil samples.

The natural antioxidants such as tocopherols are typically measured using normal phase HPLC with fluorescence detection. Typical separations of tocopherol isomers present in selected oils are presented in Figure 11.3. The tocopherol isomers were quantified using calibration for each individual isomer. The total amounts and the amount of individual isomers of tocopherols are presented in Table 11.2. Changes of tocopherol content are evident by examining oils with different degrees of oxidative degradation (Figure 11.4). Measurement of the amount of tocopherols further complements the description of an oil oxidation status. The need to measure degradation of individual tocopherol isomers is related to their different oxidative degradation rate.

Graduate students identify the origin of oils using the fatty acid and tocopherol composition as the main base, and compare them to an available database of the compositions of oils and fats. The final step in the analysis for the graduate students involves a statistical assessment of accuracy and reproducibility of the applied analytical procedures.

11.4 FINDING AND DESCRIBING RELATIONSHIPS

After completion and tabulation of the results, the nutritional significance of these findings can be assessed and related to the measured parameters and the nutritional quality of the oils. Students are required to prepare a report relating current knowledge of human nutrition to the results obtained in these exercises. In the report, students are asked to address the following topics:

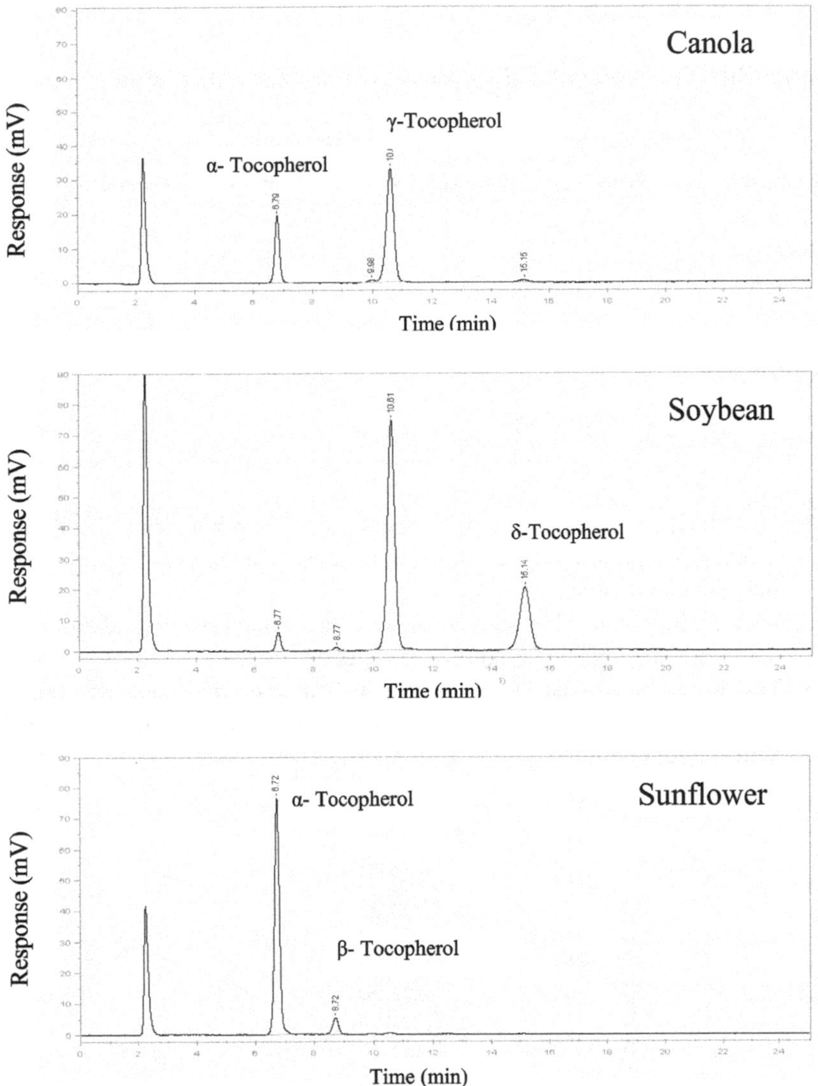

FIGURE 11.3 Separation of tocopherol isomers present in selected oils.

1. Provide a brief description and discuss the bases of the analytical procedures applied.
2. Explain the information each measured parameter provides and what it means in oil.
3. Discuss the relationship between fatty acid composition and nutritional quality of oil.
4. Discuss how the composition of vegetable oil can be used to establish the origin of the oils.
5. What does the tocopherol content and composition tell you about the oil?

Composition and Content of Tocopherols in Selected Oils (ppm)

Tocopherol	Oils			
	Palm Olein	Canola	Mid Oleic Sunflower	Hydrogenated Canola
α-Tocopherol	180	289	786	276
β-Tocopherol			18	
γ-Tocopherol		486	46	512
δ-Tocopherol				
Others*	1020			
Total Tocopherols	1200	775	850	788

*Others—include isomers of tocotrienol

TABLE 11.2

6. Describe the relationship between the presence of peroxides and carbonyls as measured by peroxide and anisidine values, respectively, and oil quality and nutritional value.
7. What is the relationship among sensory data, oxidative degradation measurements and nutritional quality of vegetable oils?
8. How would you translate compositional data and other measured parameters into nutritional information/recommendation for consumers? Substantiate your approach.

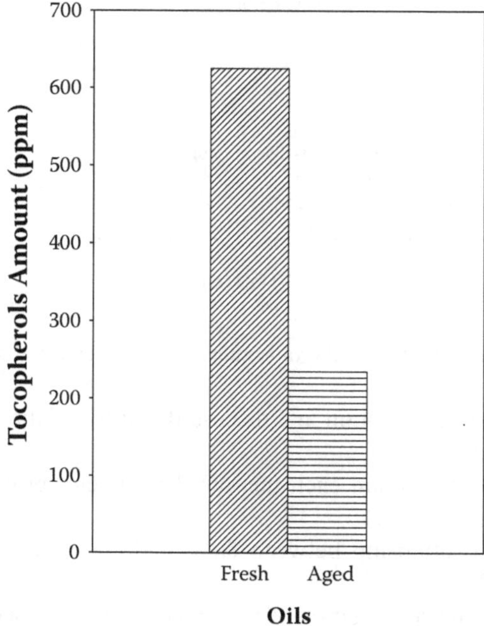

FIGURE 11.4 Changes in tocopherol during aging of canola oil.

Addressing these topics requires a detailed explanation of the analytical results produced during the assessment of the oils and their nutritional relevance. In developing the report, students need to discuss the relationship between the parameters assessed using the analytical procedures and the nutritional impact of these findings. Students are also encouraged to develop conclusions in the form of recommendations for consumers that take into consideration the relevant information acquired in the quality assessment exercises.

In the report, the relationship between the fatty acid compositions and human health needs to be discussed, particularly as it applies to the amount of the essential fatty acids.[4] Saturated, monounsaturated and polyunsaturated fatty acids all affect blood cholesterol levels and development of heart and other diseases differently.[5]

Treating all polyunsaturated fatty acids the same is incorrect. There are two well-recognized groups of essential fatty acids, n-6 and n-3; each of them has a different metabolic effect on the human body. Recommended ratios between these two groups of PUFAs (n-6/n-3) should be within 3 to 5 to assure proper balance among metabolites produced from each group. It is well documented that both groups of fatty acids are metabolized by elongation and desaturation, utilizing the same enzymes to produce important eicosanoid and prostaglandins. The later metabolites are involved in the regulation of many physiological functions in the human body.[4,5] An improper balance of these fatty acids will cause a deficiency of one of the eicosanoids, resulting in physiological and biochemical abnormalities in humans.

The North American diet is generally overloaded with n-6 fatty acids, particularly linoleic acid, which is the main PUFA in most commercial oils and fats used in food formulation and preparation.[5] Avoidance of omega-3 fatty acids, mainly linolenic acid, is related to the fact that linolenic acid oxidizes at about twice the rate of linoleic acid, causing earlier formation of rancidity and lower stability of food. Formulators generally avoid addition of oils with linolenic acid content higher than 2% in foods that have led to a deficiency of this group of essential fatty acids in our diets.[5] In addition to fatty acids, this exercise also includes analysis of tocopherols, natural antioxidants mainly produced by plants, which serve as the most efficient antioxidants available to humans.[4] Nutritional considerations of vegetable oils need to include the content and type of natural antioxidants present in oils to better describe their nutritional properties. Based on the results of oil composition, the student is encouraged to formulate nutritional recommendations.

Oxidation can greatly affect the quality of a vegetable oil, which is evidenced by a reduction in the amounts of polyunsaturated fatty acids and tocopherols, the production of oxidation products and the formation of off-flavor components. It has been well documented that oxidation of unsaturated fatty acids can produce components that have detrimental effects on human health. Measuring the presence and the amount of oxidative degradation components adds information about the quality of oils and their applicability as food, ingredient in food formulation and medium for food preparation. Analyses of off-flavor components, peroxide and anisidine value describe the status of a vegetable oil, thus facilitating the choice of which fat or oil to be used in food applications.

In addition to all analyses for undergraduates, a more in-depth analysis and interpretation is expected from graduate students. They are expected to identify the origin of the oil and fat samples provided, perform statistical analysis using the analytical data obtained and critically review the analytical methods applied.

Identification of an oil origin is based on the fatty acid and tocopherol content and composition. Each sample of oil has a characteristic composition of fatty acids and tocopherols, utilizing available databases of composition of oils; it is possible to establish the origin of oils with confidence, assuming the oil is not extensively oxidized or modified. Recent developments in plant breeding and genetic modification have altered the fatty acid profile of a number of commercial oils, making their identification based solely on the fatty acid composition more difficult. Consequently, other factors such as the tocopherol or sterol content and composition need to be considered for origin identification that is usually less affected by the modification.

Statistical evaluation of analytical procedure is based on examination of cumulative data produced by all students for particular oils. The exercise is designed in such a way that four different students or operators assess the same oil sample. Utilizing at least four replications for all oils analyzed, basic statistical analysis can be performed to assess the accuracy and reproducibility for each analytical procedure used in the exercise. Following statistical analysis, graduate students are asked to critically assess the methods for validity and potential problems in performing procedure. The graduate students are also encouraged to suggest improved methods of these analyses and how these new approaches will offer a better characterization of fats and oils.

The described exercises provide the students with hands-on experience when performing analyses as prescribed by approved analytical procedures. The most valuable experience for students is in the interpretation of results and finding relationships of how measured values obtained using the analytical methods relate to human metabolism and health. This process allows students to formulate nutritional recommendations, and describe and determine characteristics of vegetable oils that are beneficial to human nutrition. Some of the analytical processes are more complex and will need some specialized equipment. A good knowledge of nutrition and analytical chemistry is required to recognize the evidence one obtains from these exercises, to more fully assess the quality of oils used in food products and to appreciate the affects on human nutrition and health.

ACKNOWLEDGMENT

Thanks to Dr. John K.G. Kramer for his critical evaluation of the manuscript.

REFERENCES

1. AOCS method Ce 1h-05. *Official and Recommended Practices of the American Oil Chemists Society*, 5th ed., AOCS Press, Champaign, IL, 1997.
2. *Official and Recommended Practices of the American Oil Chemists Society*, 5th ed., AOCS Press, Champaign, IL, 1997.

3. Przybylski, R., Development of a trapping system for isolation of volatile flavor compounds from different food matrices, in *Food Flavors and Chemistry, Advances of the New Millennium*, Spanier, A.M. et al., eds., p. 373. Royal Society of Chemistry, Cambridge, 2001.
4. Gebauer, S., Harris, W.S., Kris-Etherton, P.M., and Etherton, T.D., Dietary n-6/n-3 fatty acid ratio and health, in *Healthful Lipids*, Akoh, C.C. and Lai, O.-M., eds., pp. 221–248, AOCS Press, Champaign, IL, 2005.
5. Whitney, E. and Rolfes, S.R., The lipids: Triglycerides, phospholipids, and sterols, in *Understanding Nutrition*, 10th ed., chap. 5. Thomson & Wadsworth, Belmont, 2005.

12 High-Temperature Gas–Liquid Chromatographic Profiling of Plasma Lipids: A Student Exercise

Arnis Kuksis

CONTENTS

12.1 INTRODUCTION

The analysis of blood lipids constitutes an important adjunct to clinical diagnosis of hypercholesterolemia and hypertriglyceridemia, which are believed to be important risk factors for heart disease and stroke.[1] Various chemical and enzymatic methods have been utilized in the determination of plasma lipids, some of which have been automated and are currently widely applied.[2] The development of direct gas–liquid chromatography (GLC) analysis of total lipid extracts of plasma or serum is based on a successful chromatography of the individual components of the plasma total lipid mixture.[3] Due to the advantageous distribution of the molecular weights and functional groups of the neutral lipid moieties, the plasma lipids are especially well suited for programmed temperature resolution on nonpolar GLC columns. For optimum resolution and recovery, the lipid extracts of whole plasma or individual lipoprotein classes are subjected to dephosphorylation and trimethylsilylation (TMS) prior to GLC. Specifically, the GLC resolution provides a quantitative estimate of plasma cholesterol, triacylglycerols and phospholipids, along with the degree of unsaturation of the major fatty acid moieties. This approach to the quantitative GLC analysis of plasma lipids, which was originally proposed over 30 years ago,[4] has been extensively utilized in plasma lipid analyses in health and disease, and the results have been reviewed.[5]

High temperature GLC profiling of plasma lipids was introduced in the Advanced Biochemistry Laboratory at the University of Toronto some 35 years ago as a means of stimulating student interest in lipid metabolism and in lipid research. A rapid high temperature GLC resolution and quantification of all major plasma lipid classes was initially performed on each student's own blood sample and secured undivided interest of the class, although at some loss of privacy of the clinical information. Later, when individual blood letting was discontinued, the exercise was continued using either commercial plasma samples or outdated plasma samples supplied by a local hospital.

This chapter provides a detailed account of the methodology, typical results obtained and some personal observations made over a 30-year period of teaching plasma lipid profiling to the Advanced Biochemistry Laboratory at the University of Toronto.

12.2 METHODS AND MATERIALS

The high temperature profiling of intact plasma lipids by GLC was based on an early discovery of a successful molecular weight resolution of natural triacylglycerols.[6] Several factors have since been recognized as critical for effective high molecular weight lipid ester profiling, including a short column, nonpolar liquid phase, on-column injection and temperature programming. Important refinements constituted optimal sample load, flame ionization detection and nature of carrier gas.[7]

12.2.1 Gas Chromatographic System

The original experiments were performed with a Hewlett-Packard Model 5700 Automatic Gas Chromatograph, equipped with a short packed stainless steel column and a hydrogen flame ionization detector (Hewlett-Packard, Palo Alto, CA). In parallel, a Varian Model 2700 Moduline gas chromatograph (Varian Instruments, Bremen, Germany) coupled to a Varian MAT CH-5 single focusing mass spectrometer was utilized for peak identification. A 60 cm × 2 mm ID glass column packed with 3% OV-1 on Gas Chrom Q was substituted for the stainless steel tube. Eventually, a Hewlett-Packard Model 5880 Capillary Gas Chromatograph equipped with a short flexible quartz column was used for demonstration of the potential of capillary GLC for plasma lipid profiling.

12.2.1.1 Column Selection

Initially, short (50 cm x 2 mm ID) stainless steel columns were used for durability. The columns were filled in the laboratory with a porous support consisting of a flux-calcined diatomaceous earth (100–120 mesh), which had been rendered hydrophobic with trimethylchlorosilane and then coated with a thin film (1–3%, w/w) of a methylsiloxane polymer. Such packings later became commercially available under a variety of trade names, including SE-30, JXR, OV-1 on Gas Chrom Q (Applied Science Labs., Inc.) or Supelcoport (Supelco Inc.) or equivalent preparations. The columns were connected directly to 1/8 in. unheated on-column injector and 1/8 in. detector connectors.

Following overnight heating at maximum operating temperature under full flow of carrier gas, these columns possessed low vapor pressure and gave a stable baseline in the 200 to 340°C temperature range. Other methyl silicone polymers performed similarly, although the time of conditioning varied. Longer column lengths decreased recovery of the higher molecular weight components, while shorter column lengths impaired resolution of peaks. Only minor improvement in the peak shape and recovery was observed by substituting glass for stainless steel as column material.

Eventually, plasma total lipid profiling (as a demonstration) was performed on short lengths (8 m) wide-bore (0.32 mm ID) flexible quartz capillary columns coated with chemically bonded SE-54 methylphenylsilicone liquid phase (Hewlett-Packard, Model 5880 Gas Chromatograph). The longer, more efficient columns gave further separation by carbon number, which practically eliminated lipid class overlaps.

12.2.1.2 Choice of Carrier Gas

The low resolution of the lipid classes possible on the short packed columns can be readily achieved with nitrogen as a carrier gas, although helium has been shown to provide notable improvement for the separation and recovery of the higher molecular weight triacylglycerols. Hydrogen is better suited for the longer packed columns and capillary columns. The use of hydrogen as a carrier gas requires precaution, as it constitutes an explosion hazard in a confined area. However, the use of hydrogen flame ionization detection, which is absolutely necessary for lipid profiling, also releases hydrogen in the environment. Elution times decrease and peak recovery increases with increasing flow rate. However, separation efficiency may decrease.

12.2.1.3 Temperature Programming

Prior to use, the columns were conditioned at 350°C for 3 hours. The carrier gas (N_2) flow was 40 ml/min with a hydrogen flow adjusted to 30 ml/min and the air flow set at 240 ml/min. The GLC separations were routinely made by temperature programming from 175–350°C at 4°C/min or 8°C/min, with the columns in a dual compensating mode. For capillary GLC on nonpolar columns, stepwise temperature programming was best, for example, 40°C (isothermal for 5 min); then 30°C/min to 150°C; then 20°C/min to 230°C; then 10°C/min to 280°C; then 50°C/min to 340°C; then holding to the end of the run. For GLC on polar capillary columns, the temperature program was modified further: 40°C–290°C, ballistic; 290°C, isothermal for 5 min; 290–330°C, 10°C/min; 330–360°C, 5°C/min.

12.2.2 Lipid Standards

The plasma lipid profiling was standardized by analyzing a synthetic sample of plasma lipids obtained by combining free fatty acids (FFAs), monoacylglycerols, free cholesterol, diacylglycerols, ceramides, cholesteryl esters and triacylglycerols in the proportions in which they are known to occur in normolipemic plasma. In addition, real plasma samples of known content of free and total cholesterol and triacylglycerols (Reference Laboratory of the Lipid Research Clinics Program, Atlanta, GA; or a commercial supplier) served as secondary controls. Both types of standards were resolved and quantified in the presence of the internal standard tridecanoylglycerol added at 10% level of the total lipid prior to sample digestion with phospholipase C and trimethylsilylation.

12.2.1 Synthetic Plasma Lipid Standard

The synthetic plasma lipid samples were prepared by weighing out a minimum of 10 mg of each lipid and dissolving it in 100 mL of chloroform in a screw cap volumetric flask and combining appropriate volumes of not less than 1 mL each to give the desired proportions. The absolute concentration of each lipid ranged from 0.1 to 100 μg/μL of the final solution.

12.2.2 Selection of Internal Standard

The internal standard was selected to approximate the physical and chemical properties of the major lipid classes in the sample and possessing a chromatographic retention time well within the midrange of the elution times of the major plasma lipid components. These conditions were met by both tridecanoylglycerol and cholesteryl acetate, neither of which tended to overlap with the common plasma lipid components. Tridecanoylglycerol was selected as the standard because of stability and availability in high condition of purity. It was diluted to 100 μg/mL in chloroform. The absolute concentration of the tridecanoylglycerol internal standard was maintained at 1 μg/μL.

12.2.2.3 Analyses of Fatty Acids

Aliquots of a Folch extract of plasma total lipids were separated into lipid classes by preparative thin-layer chromatography (TLC) and each lipid fraction was scraped from the plate and transesterified in 3N methanolic HCl in a sealed vial under nitrogen atmosphere at 100°C for 45 min.[4] The resulting fatty acid methyl esters were extracted with hexane and the extracts reduced to small volume for GLC. The fatty acid methyl esters were quantified by GLC using a polar liquid phase. The quantitative results were compared with published fatty acid composition of plasma neutral lipids and phospholipids, and the significance of the findings discussed.

12.2.3 Total Plasma Lipids

For the present purposes, total plasma lipids consist of FFAs, neutral lipids and the neutral lipid moieties of the plasma glycerophospholipids and sphingomyelins. When expressed on a mol% basis, these compounds provide true molar proportions of the total plasma lipids. It is recognized, however, that free diacylglycerols and free ceramides occasionally found in plasma in detectable amounts could overlap with the diacylglycerols and ceramides released from phospholipids by phospholipase C and thus lead to an overestimation of the plasma phospholipid content.

12.2.3.1 Enzymatic Dephosphorylation

Ethylene diamine tetraacetate (EDTA) (0.01%) plasma or serum (0.1–1.0 mL) was added to a screw-cap centrifuge tube (18 mL capacity) containing 0.2–0.4 mg phospholipase C (α-toxin of *Clostridium welchii*, Sigma Chemical Co., St. Louis, MO) in 4 mL of 17.5 mM tris buffer pH 7.3 along with 1.3 mL of 1% $CaCl_2$ and 1 mL of diethyl ether, and the mixture incubated with shaking for 2 hours at 30°C.[4] To the reaction mixture were then added 5 drops of 0.1 N HCl and it was extracted once with 10 mL of chloroform/methanol 2:1 containing 150–250 µg tridecanoylglycerol as internal standard. The solvent phases were separated by centrifugation for 10 min at 200 × g. The clear chloroform phase was removed from the bottom of the tube and was dried by passing through a Pasteur pipet containing 2 g of anhydrous Na_2SO_4. The effluent was evaporated under nitrogen and the residue subjected to trimethylsilylation.

12.2.3.2 Zeolite Adsorption

Alternatively, the plasma samples were dephosphorylated by passing the chloroform solution of a Folch extraction of the plasma sample through a Zeolite or another silicic acid adsorbent column, which retained the phospholipids while allowing the neutral lipids, as well as the FFAs, to pass through.[8] The effluent was collected, diluted with internal standard and saved for trimethylsilylation. In this instance the plasma total lipid profile would exclude the phospholipids, the determination of which is not necessary for a simple clinical diagnosis of hypercholesterolemia and hypertriglyceridemia.

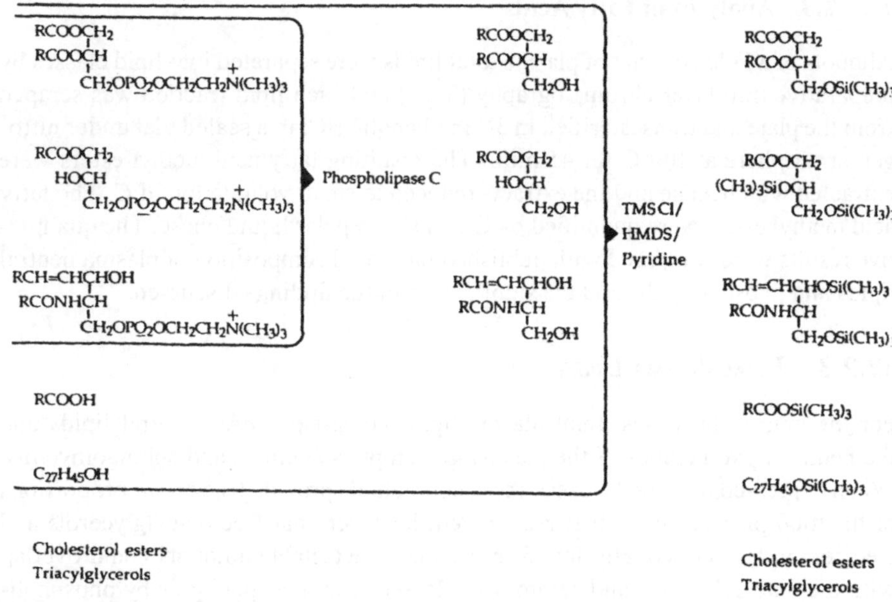

SCHEME 12.1 Summary of sample preparation. TMSCl, trimethylchlorosilane; HMDS, hexamethyldisilazane.

12.2.3.3 Trimethylsilylation

For quantitative gas chromatographic analyses, the free hydroxyl groups of the sterols, mono- and di-acylglycerols, and ceramides were converted into TMS ethers and the carboxy groups of the FFAs into TMS esters[4]. For trimethylsilylation, appropriate volumes of the standards and unknowns were evaporated under nitrogen and the residues dissolved in 150–250 μL of hexamethyldisilazane/trimethylchlorosilane/pyridine (12:5:2, by vol) and transferred to a sampling vial and the vial sealed. After 30 min at room temperature the silylation was complete and the samples ready for GLC. The sample work-up is summarized in Scheme 12.1.

12.3 TOTAL LIPID PROFILING

2.3.1 Resolution of Synthetic Standards

Analysis of synthetic standards served to introduce the method to the student class and to establish the optimum conditions of peak separation and recovery, as well as to acquaint the class with the principles of peak identification and quantification. Figure 12.1 shows a GLC separation of neutral lipids and FFAs on a nonpolar packed column of short length.[4] It provides an excellent separation of a synthetic mixture of lipids designed to mimic plasma lipids after digestion with phospholipase C (A). The separation of an equal weight mixture of synthetic triacylgycerols and free cholesterol provides a simple reference standard for completeness of trimethylsilylation of cholesterol and the recoveries of the high molecular weight components (B).

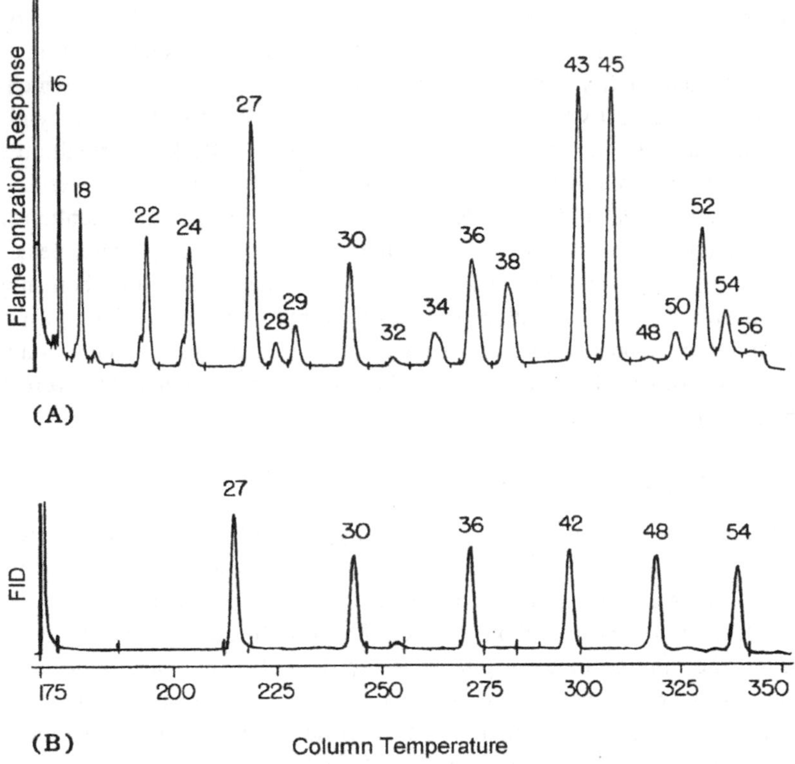

FIGURE 12.1 GLC of neutral lipids and FFAs on nonpolar packed columns of short length.[9] (A) A synthetic mixture of lipids designed to mimic plasma lipids after digestion with phospholipase C. (B) An equal weight mixture of synthetic triacylglycerols and free cholesterol. Peak identification: 16 and 18, TMS esters of C_{16} and C_{18} fatty acids; 22 and 24, TMS ethers of monoacylglycerols of C_{16} and C_{18} fatty acids; 27–29, TMS ethers of cholesterol, campesterol and sitosterol; 30, tridecanoylglycerol (internal standard); 34–38, TMS ethers of dipalmitoyl, palmitoylstearoyl and distearoylglycerols; 43 and 45, cholesteryl palmitate and stearate; 48–54, triacylglycerols of 48 to 54 total acyl carbons per molecule. Instrument: Hewlett-Packard Model 5700 Automatic Gas Chromatograph equipped with a flame ionization detector, unheated on-column injector, dual columns, differential electrometer, electronic peak area integrator and an automatic sample injector. Columns: stainless steel tubes, 50 cm × 2 mm ID, packed with 3% OV-1 on Gas Chrom Q. Carrier gas, nitrogen at 80 ml/min. Detector, 350°C. Temperature program: 4°C/min from 175 to 350°C.

The peaks in (A) are identified in order of their elution as follows: Peaks 16 and 18, TMS esters of the C_{16} and C_{18} carbon fatty acids; Peaks 22 and 24, di-TMS ethers of C_{16} and C_{18} monoacylglycerols; Peaks 27, 28 and 29, TMS ethers of cholesterol, campesterol and sitosterol; Peak 30, tridecanoylglycerol (internal standard); Peaks 34, 36 and 38, TMS ethers of dipalmitoyl-, palmitoylstearoyl- and distearoylglycerols; Peaks 43 and 45, cholesteryl palmitate and cholesteryl stearate; Peaks 48 to 54, triacylglycerols with 48 to 54 acyl carbons per molecule.

The peaks in (B) are identified as follows: Peak 27, free cholesterol, which emerges slightly ahead of the TMS ether of cholesterol; Peak 30, tridecanoylglycerol internal standard, the elution time of which coincides exactly with that of the internal standard in (A); Peak 36, trilauroylglycerol, the elution time of which corresponds to that of the TMS ether of distearoylglycerol; Peak 42, trimyristoylglycerol, which migrates slightly ahead of cholesteryl palmitate; Peak 48, tripalmitoylglycerol, the elution time of which corresponds closely to that of the mixed acid C_{48} triacylglycerol peak in (A); and Peak 54, tristearoylglycerol, the retention time of which is slightly longer than that of the mixed acid C_{54} triacylglycerol peak in (A). The nearly equal height triacylglycerol peaks in (B) indicate that both high and low molecular weight triacylglycerols are recovered in the proportion in which they exist in the injection solution. This suggests that the various lipid classes analyzed in (A) may also have been recovered in the proportions in which they were present in the injection solution.

12.3.1.1 Peak Identification

The peak identification in gas chromatography is usually based on the relative retention times, although absolute retention times may also be utilized. From the nonpolar liquid phase, the TMS derivatives of the neutral lipids are eluted in order of increasing molecular weight or carbon number. The unsaturated esters migrate slightly ahead of the more saturated esters of the same carbon number. Under controlled conditions, the retention times are highly reproducible when measured in relation to an internal standard. The absolute retention times, however, decrease with increasing use of the column due to gradual loss of the liquid phase.

12.3.1.2 Peak Quantification

Table 12.1 indicates the reproducibility of the relative retention times and measurements of peak areas in relation to internal standard as obtained by a flame ionization detector, the signal of which is known to correspond closely to the carbon content of the analyzed molecules.[4] The peaks may be quantified from the areas under the curve by means of external or internal standardization. In external standardization, the peak area response is plotted in response to increasing concentrations of the injected lipid component. In the case of the hydrogen flame ionization detector, a linear response is readily obtained over a very wide range of concentrations. In practical chromatography, the use of an internal standard is preferred, as it does not require accurate dilutions and injections. Every component can be measured in relation to the known weight of another component (internal standard) added as a reference. It makes no difference how many injections are made; it can be assumed that the entire sample along with all internal standard is injected every time.

The weight of the total mixture can be obtained by summing the areas of the individual components as previously described.[4] All components are calculated in terms of mg% according to Equation 12.1:

$$W(X) = [A(X)/A(IS)]\ W(IS).F(WX), \qquad 12.1$$

where F(WX) is the weight response correction factor of component X, W(X) and W(IS) are weight % for component X and the internal standard, respectively. A(X) and A(IS) are peak areas for component X and the internal standard, respectively. The weight response factor F(WX) for component X is inversely proportional to the relative response and can be defined by Equation 12.2:

$$F(WX) = [W(X)/W(IS)]\,[A(IS)/A(X)], \qquad 12.2$$

where the variables are defined as in Equation 1. Similarly a molar response correction factor F(MX) may be defined by Equation 3:

$$F(MX) = [M(X)/M(IS)]\,[A(IS)/A(X)], \qquad 12.3$$

where M(X) and M(IS) are the numbers of mole % for component X and the internal standard, respectively, while A(IS) and A(X) are again the peak areas of compounds X and the internal standard.

When related to the known weight of the internal standard added at about 10% of the total sample weight, accurate estimates are obtained for each and all components of the mixture. Because slight differences are known to occur among the

TABLE 12.1
Reproducibility of Relative Retention Times and Measurement of Peak Areas by Internal Standard for a Synthetic Mixture at Optimum Operating Conditions *

Component	Relative Retention Time	Peak Area
Palmitic acid	0.053 ± 0.0015	4.325 ± 0.139
Stearic acid	0.099 ± 0.0010	1.708 ± 0.006
Monopalmitin	0.308 ± 0.0031	6.805 ± 0.102
Monostearin	0.485 ± 0.0057	3.320 ± 0.054
Cholesterol	0.703 ± 0.0052	8.468 ± 0.067
Campesterol	0.779 ± 0.0037	1.843 ± 0.013
Sitosterol	0.842 ± 0.0033	3.099 ± 0.020
Tridecanoin	1.000	2.692 ± 0.039
Dipalmitin	1.259 ± 0.0041	5.348 ± 0.040
Palmitostearin	1.377 ± 0.0052	8.775 ± 0.061
Distearin	1.478 ± 0.0064	9.823 ± 0.114
Cholesteryl palmitate	1.734 ± 0.0085	6.678 ± 0.026
Cholesteryl stearate	1.852 ± 0.0095	11.090 ± 0.057
Dipalmitostearin	2.095 ± 0.0114	2.615 ± 0.027
Distearopalmitin	2.185 ± 0.0175	13.945 ± 0.095
Tristearin	2.269 ± 0.0124	5.006 ± 0.092

*Data from five consecutive runs. Reference peak retention time was 1016 seconds. Tridecanoylglycerol was the reference standard.

flame ionization responses of different lipid components, accurate work requires the determination of peak area correction factors.[9]

12.3.1.3 Correction Factors

The true mass of the eluted components was determined in relation to that of the internal standard by multiplying the experimentally observed peak area ratios by the reciprocal of the ratio of the weights of the test component and the internal standard. Table 12.2 lists the peak area correction factors for the common components of the neutral lipids derived from plasma lipids.[4] The highest correction factors must be applied to the TMS derivatives of the fatty acids, monoacylglycerols and cholesterol, while the TMS ethers of diacylglycerols and ceramides yield a response comparable to that of the cholesteryl esters and triacylglycerols. These factors apply to all lipids in the linear ranges indicated, as well as in the adjacent lower and higher ranges of the more volatile compounds. Special correction factors were necessary for triacylglycerols below 2 μg per peak. Larger peaks preceding smaller ones were observed to exert a significant priming effect upon recovery of the smaller ones.

12.3.2 Resolution of Plasma Lipids

The unknown plasma samples were obtained from the students' own blood, then analyzed along with plasma samples obtained from normal subjects on specific diets and patients with hypercholesterolemia and hypertriglyceridemia. For the purposes of peak identification and quantification, the chromatograms illustrating the effect of diet and disease have been selected.

TABLE 12.2
Peak Area Correction Factors for Various Neutral Lipids Under Optimum Chromatographic Conditions (> 2 μg/peak)

Plasma Lipid or Derivative*	Working Range (g/peak)**	F(W)(S.D.)***
Fatty acid TMS ether	0.16–1.27	0.75 ± 0.013
Monoglyceride TMS ether	0.63–5.5	0.71 ± 0.007
Cholesterol TMS ether	0.127–50.8	0.71 ± 0.007
Diglyceride TMS ether	0.15–3.0	0.91 ± 0.01
Cholesteryl ester	1.3 –12.7	0.85 ± 0.01
Triglyceride (lard)	4.75–38.0	0.98 ± 0.012
Tridecanoin	1.0–25.0	1.00
Ceramide TMS ether	As for diglycerides	

*Test species included the C_{16} and C_{18} fatty acids and their glyceryl and cholesteryl esters.

**Mass ranges given in the underivatized form.

***Minimum of five injections over the mass range. All injections made from a solution of 0.15 mL TRISIL/BSA using 1.9 μL/injection. F(W) as described in text.

12.3.2.1 Effect of Dietary Fat

Figure 12.2 shows the GLC profiles of plasma total lipids as obtained on nonpolar columns for the same subject following consumption of (A) fat-free (B) corn oil and (C) butterfat (40% calories) diet for 1 week each.[10] Peak identification is shown in the chromatograms. It is obvious that the plasma lipid profile is significantly influenced by dietary fat. Consumption of corn oil results in a marked increase in the proportion of C_{54} triacylglycerols, while consumption of butterfat results in a relative increase in the shorter chain length (C_{48} andC_{50}) triacylglycerols. Consumption of a fat-free diet results in plasma triacylglycerols made up of C_{50} and C_{52} acyl carbons. The C_{50}/C_{54} ratio is highly characteristic of the dietary fat consumed. The lipid profiles of the students fell somewhere between those of fat-free and butterfat diets, especially of those who had not fasted overnight before blood taking.

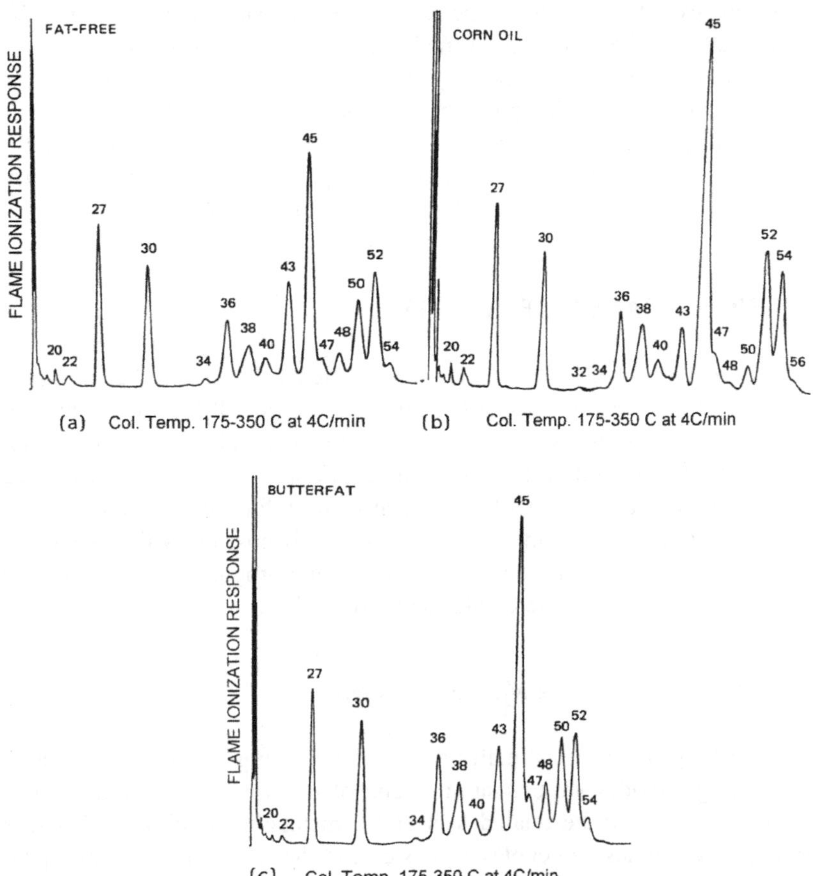

FIGURE 12.2 GLC of plasma total lipids after dephosphorylation as obtained from the same subject following consumption of (a) fat-free, (b) corn oil (40% of calories), and (c) butterfat (40% of calories) diets for 1 week each.[10] Peak identification, instrumentation and GLC conditions were as given in Figure 12.1.

12.3.2.2 Effect of Disease

GLC profiling of total lipid extracts is eminently suited for spotting lipid abnormalities in plasma or in plasma lipoproteins. A reference to a normal lipid profile recorded in the presence of an internal standard reveals the presence or absence of a general or specific hyperlipemia. Increased levels of cholesterol, with or without increases in triacylglycerols, indicate hypercholesterolemia.[1] Relative absence of cholesterol ester or an increase in free cholesterol/cholesterol ester ratio indicates lecithin-cholesterol acyltransferase (LCAT) deficiency.[1] An excess of plant sterols allows the recognition of phytosterolemia.[1] Especially easy to detect are increases in plasma FFAs and monoacylglycerols (representing lysophosphatidylcholine), suggesting increased phospholipase activity due to infection. Figure 12.3 shows the total lipid (A) and the neutral (B) profile of a hyperlipemic subject. In relation to the internal standard (Peak 30), all the lipid classes in (A) have been increased, although to a variable extent.[11] In addition, peaks for the plant sterols, campesterol (28) and sitosterol (29), can be readily discerned. The neutral lipid fraction, obtained by a Zeolite removal of the phospholipids, shows (B) that despite the large increase in triacylglycerols and phospholipids, the free diacylglycerol and ceramide contribution has remained minor and largely unchanged from that found in normolipemic subjects. Table 12.3 shows a quantitative comparison of the reproducibility and accuracy of analyses of normal and hyperlipemic plasma samples.[4]

12.4 SIGNIFICANCE OF RESULTS

Extensive analyses of plasma samples from patients showed marked alterations in the total lipid levels as well as in the lipid class ratios. It has been shown previously that the separation and quantification of the major classes of plasma lipids by automated GLC following a limited preliminary work-up is a practically feasible analytical routine. It offers an attractive opportunity for the simultaneous monitoring and study of many more metabolites than usually possible with other automated systems, which should facilitate both diagnosis and prognosis as well as treatment of all diseases reflected in altered plasma lipid profiles.

12.4.1 Correlation with Metabolic Events

In addition to quantitative estimates of each peak in mg% or mole%, plasma total lipid profiling provides various summations of the lipid classes and certain lipid class ratios, which can be related to specific metabolic conditions.[5] These ratios include the free cholesterol/cholesteryl ester and the free cholesterol/phosphatidylcholine (or total phospholipid) quotient, and the sphingomyelin/phosphatidylcholine quotient, which are of interest to clinical biochemists and medical researchers.[12,13] The ratios of the lipid classes obtained for the reference samples corresponded closely to those recorded in the literature for normal subjects or to those which could be calculated from published records.

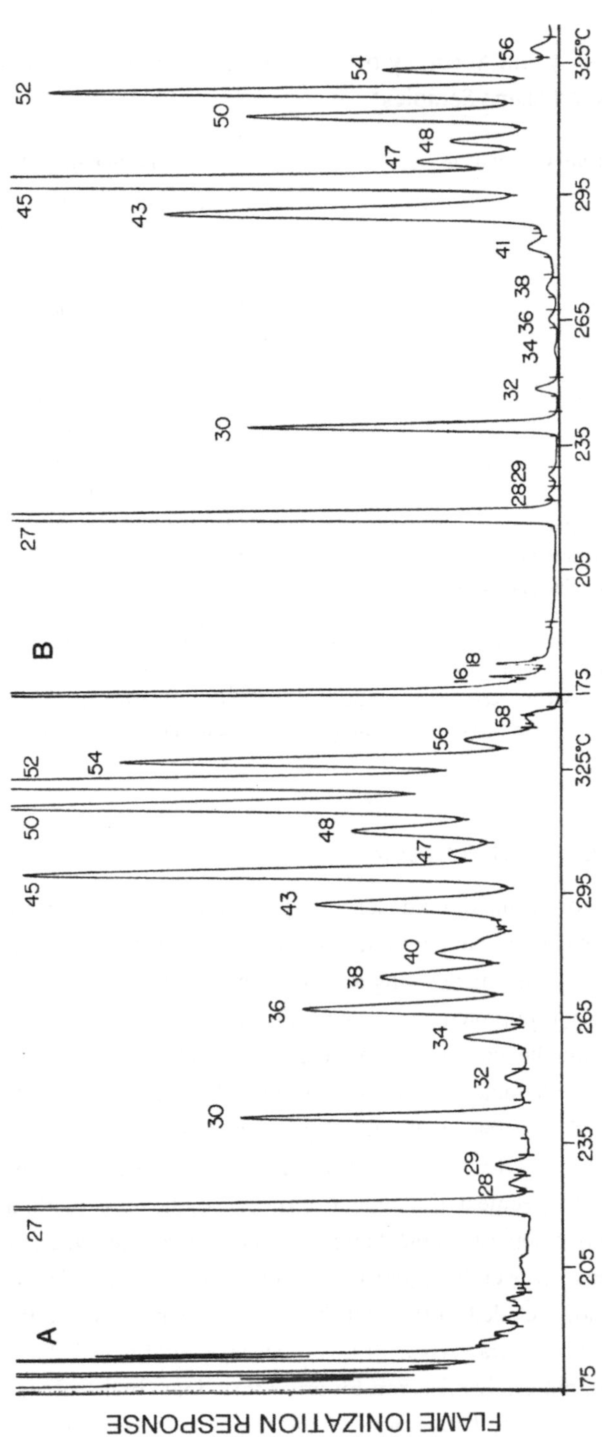

FIGURE 12.3 GLC of plasma total (A) and neutral (B) lipid extracts from two hyperlipemic subjects. Peak identification, instrumentation and GLC conditions were as given in Figure 12.1.

TABLE 12.3
Reproducibility and Accuracy of Analyses of Normolipemic and Hyperlipemic Plasma Samples*

Lipid Components	Concentration (mg%)**	
	Normolipemic	Hyperlipemic
FFAs	16.1 ± 1.0	5.1 ± 0.5
Monoglycerides	3.9 ± 1.9	1.5 ± 0.1
Free cholesterol	39.9 ± 1.9	68.5 ± 5.9
Ceramides	34.9 ± 1.7	42.4 ± 3.3
Diglycerides	95.9 ± 7.1	125.4 ± 8.8
Cholesteryl esters	215.7 ± 9.0	198.0 ± 12.3
Triglycerides	82.3 ± 5.4	670.0 ± 47.6
Total cholesterol	169.3 ± 6.8	187.3 ± 13.2
Total lipid	488.9 ± 18.6	1110.9 ± 77.8
Free cholesterol/total cholesterol (moles/moles)	0.235 ± 0.008	0.365 ± 0.007
Total diglyceride/free cholesterol (moles/moles)	1.51 ± 0.07	1.15 ± 0.02
Diglyceride/ceramide (moles/moles)	2.70 ± 0.14	2.90 ± 0.06

*Six parallel analyses of one normolipemic and one hyperlipemic sample of plasma including separate enzyme digestion, extraction, silylation and automated GLC.

**Mean±S.D. for six duplicates. The Autoanalyzer values for total cholesterol and triglycerides were 175 mg% and 100 mg% for the normolipemic plasma, respectively.

12.4.2 Diagnosis of Hyperlipemias

The relative ratios of the major lipid classes represent the relative ratios of the major plasma lipoprotein classes, which possess well-defined lipid composition. We have made use of this constancy in devising a vector plot, which permits a rough diagnosis of different hyperlipoproteinemia types. Figure 12.4 shows the vector analysis used for interrelating the major plasma lipid classes. The hyperlipoproteinemia type is deduced from the characteristic range of angles (theta) previously established for 10 to 20 representative subjects of each type of disease. Table 12.4 compares the diagnoses derived from GLC and LRC (Lipid Research Clinics, Atlanta, GA) methods.[14] Although the agreement is excellent in this case, the GLC method frequently could not distinguish between subjects with upper normal and Type II lipoproteinemia levels, because of overall similarities in lipid class ratios. Patients with LCAT deficiency were not included in the plot, but a significant decrease in cholesteryl ester content, which is characteristic of this disease, should have led to marked distortion of the plot, not easily mistaken for any other lipid abnormality. Needless to say, the plasma samples of the majority of the young men and women analyzed in the student lab were normolipemic, including those taken without overnight fast.

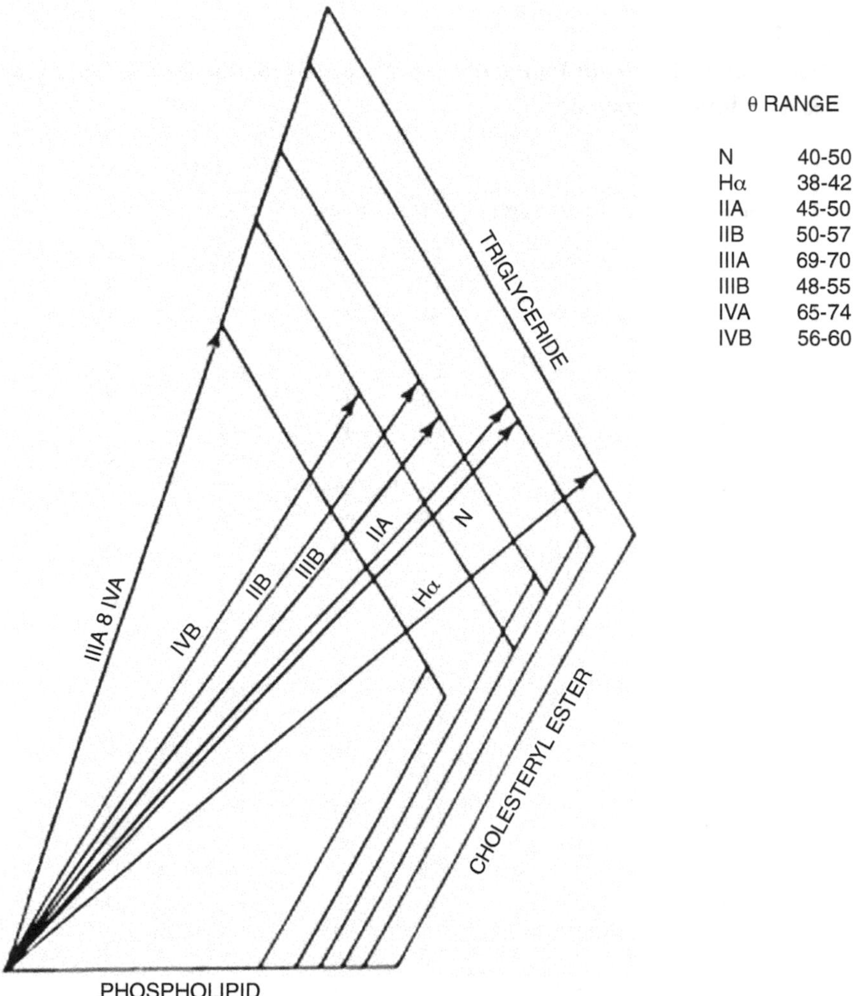

FIGURE 12.4 Vector plot constructed to interrelate the major plasma lipid classes to types of Fredrickson's hyperlipoproteinemia[1] used for class discussion.[14] N, normolipemic; Hα, hyperalphalipoproteinemia; IIA, IIB, IIIA, IIIB, IVA, and IVB, types of hyperlipoproteinemia; theta (θ), resulting angle. The plot is prepared with a minimum of footprint and maximum spread of the resulting angle. The plot lines are read from left to right in milligram%.

12.5 SUPPLEMENTARY METHODOLOGY

The student class had the opportunity to validate the methodology by gathering further supporting evidence obtained in the class by combining preliminary thin-layer chromatography with GLC/FID and by combining mass spectrometry (MS) with the TLC/GLC separations. The TLC/GLC analyses involved class participation, while the GC/MS analyses were performed as a demonstration, which the students were allowed to witness and were expected to become acquainted with by

TABLE 12. 4
Comparison of Results of Typing of Hyperlipoproteinemia by GLC and Lipid Research Clinic Methods[14]

Subject	Total Lipid (mg%)	TC/TG	Theta	Type of Hyperlipemia	
				GLC	LRC
1	718	3.8	38	Ha	N
2	654	2.0	52	N	N
3	605	2.4	45	N	N
4	656	1.7	48	N	N
5	579	2.0	44	N	N
6	860	2.4	45	N	N
7	537	1.4	47	N	N
8	601	2.9	45	N	N
9	656	2.4	43	N	N
10	1027	1.9	46	IIA	N
11	783	1.2	48	N	N
12	642	1.1	55	IIB	N
13	763	1.2	52	IIB	N
14	625	1.6	50	N	N
15	691	1.7	46	N	N
16	943	1.7	43	N	N
17	566	1.2	47	N	N
18	738	1.9	44	N	N
19	713	2.0	48	N	N
20	649	1.6	50	N	N
21	830	1.7	49	N	N
22	565	1.9	49	N	N
23	616	2.3	45	N	N
24	468	2.5	40	N	N
25	548	2.4	40	N	N
26	891	1.8	43	IIA	IIA
27	750	1.5	51	IIA	IIA
28	813	2.0	49	IIA	IIA
29	798	2.0	49	IIA	IIA
30	873	1.6	50	IIA	IIA
31	823	1.5	52	IIB	IIB
32	707	1.7	50	IIB	IIB
33	1278	0.6	68	IVA	IVA
34	1181	0.6	68	IVA	IVA
35	617	1.2	54	IIIB	IV
36	605	1.2	60	IVB	IV
37	514	3.8	38	Ha	Ha
38	992	3.2	41	Ha	Ha

Theta values as obtained from Figure 12.4. N, normolipemia; Ha, hyperalpha-lipoproteinemia; IIA, IIB, IIIB, IV, IVA, and IVB, types of hyperlipoproteinemia. Other observations are as explained in the text.

independent interpretation of the mass spectra and mechanism of electron impact ionization.

12.5.1 TLC/GLC Analyses

The TLC analyses were performed on conventional silica gel H plates (20 × 20 cm, 250 μm thick gel layer). The plates were developed with heptane/isopropyl ether/acetic acid (60:40:4, by vol) as the solvent, which retained the polar phospholipids at the origin of the plate, while the neutral lipids became resolved according to their polarity into free fatty acid, cholesterol, triacylglycerol and cholesteryl ester subclasses. Each of the fractions was scraped off the plate, recovered by extraction with chloroform and trimethylsilylated for GLC. The retention times of the resulting GLC peaks were correlated with those present in the total plasma lipid profile. In parallel, the phospholipid fraction recovered from the origin of the plate was subjected to hydrolysis with phospholipase C and the released diacylglycerols and ceramides trimethylsilylated for GLC. The retention times of the resulting GLC peaks were correlated with those present in the total plasma lipid profile and appropriate conclusions made about the identity of the plasma lipid subfractions. In addition, the total plasma lipid extract and the subfractions recovered from the TLC separation were subjected to transmethylation and the fatty acid methyl esters separated, identified and quantified by GLC on polar capillary columns. The results of the fatty acid analyses provided detailed information about the molecular species composition of the phospholipid, cholesteryl ester and triacylglycerol fractions and their likely carbon numbers.

12.5.2 GC/MS Analyses

As a demonstration, the students in the more recent classes were allowed to acquaint themselves with the results of a combined GC/MS examination of the plasma total lipids in a research laboratory. For this purpose, the dephosphorylated and trimethylsilylated plasma samples were admitted to a GC/MS instrument equipped with a nonpolar capillary GLC column, which yielded total lipid profiles very similar to those obtained by GLC/FID. Figure 12.5 shows a total electron impact ionization (EI) current profile of a plasma lipid sample (A) along with the single ion mass chromatograms representative of selected subfractions of plasma lipids, mainly free cholesterol, cholesteryl esters, tridecanoin and the ceramides arising from sphingomyelin.[15] It is seen that the ceramide peaks are readily recognized despite an extensive overlap with the diacylglycerols arising from the dephosphorylation of phosphatidylcholine. The fragment ions are identified in the legend of Figure 12.5.

12.5.3 Polar Capillary Column GLC

An improved method for determination of total plasma lipid profiles using capillary GLC was developed based on an 8 m flexible quartz capillary column coated with a chemically bonded SE-54 methylphenylsilicone liquid phase. The longer and more efficient column gave further separation by carbon number, which practically elimi-

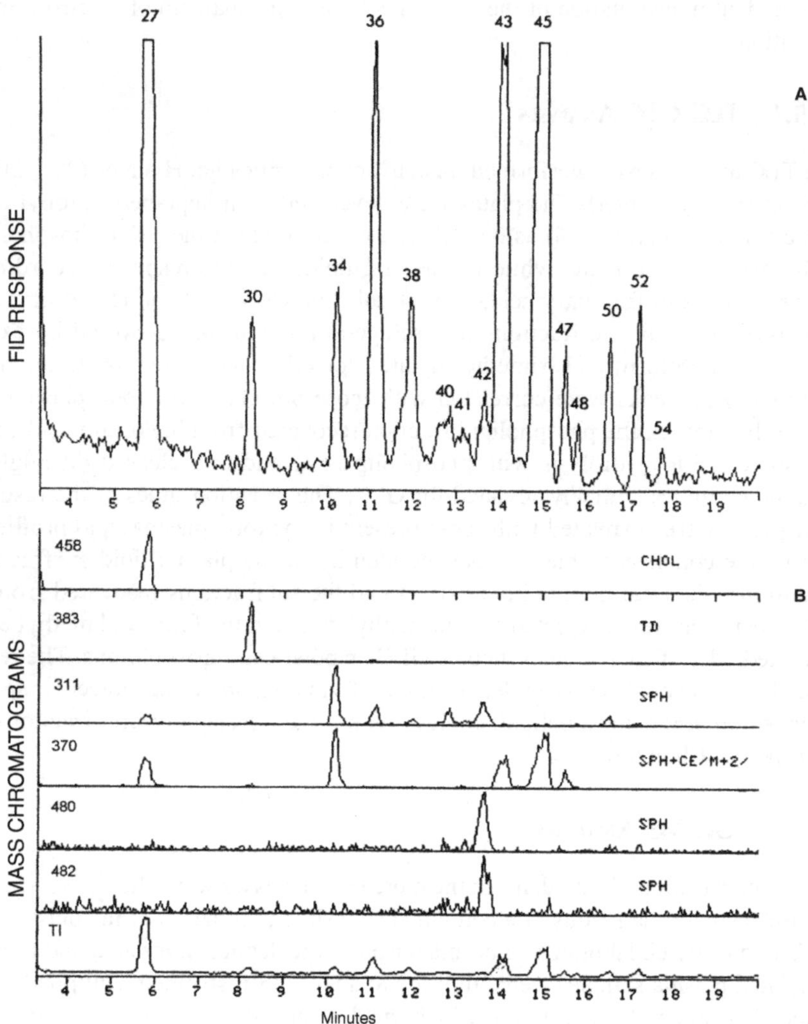

FIGURE 12.5 GC/MS profile of plasma total lipids as obtained on a nonpolar capillary column.[15] (A), total ion current; (B), mass fragmentograms of ions characteristic of ceramides (TMS ethers-CHCH$_2$OTMS)$^+$: 311, sphingosine base; 370, palmitoylsphingosine; 480, tetracosanoylsphingosine; 482, lignocerylsphingosine; 458, TMS of free cholesterol; 383, diacylglycerol-like moiety of tridecanoylglycerol; TI, total ion current. GLC conditions: Hewlett-Packard, Model 5880 Gas Chromatograph equipped with dedicated capillary inlet system (fused silica needle) and a nonpolar capillary column (8 m × 0.32 mm ID flexible quartz capillary column coated with chemically bonded SE-54 methylphenylsilicone liquid phase. Carrier gas, hydrogen at 6 psi head pressure. Mass spectrometer: Hewlett-Packard Model 5985 single quadrupole mass spectrometer connected to a capillary column via a 60 cm × 0.2 mm ID deactivated silica capillary and Supelco Supeltex M-2 (0.4/0.5 mm ID butt connector). The fused silica capillary was sealed in place at the entrance of a 1/16 in. stainless steel tubing using graphitized Vespel ferrules and an SGE connector. The interface and ion source temperatures were 350°C and 250°C, respectively. Scanning was made in electron impact mode between 290 and 890 amu in a cyclic manner.

nated lipid class overlaps. Furthermore, the use of on-column injection eliminated the discrimination against higher molecular weight components. Increasing the column length resulted in further increases in resolution, but this advantage had to be balanced against the disadvantage of decreased recoveries of the higher molecular weight components, longer peak retention times and higher final temperatures. Cholesteryl esters undergo noticeable degradation at temperatures above 310°C. Likewise, the recoveries of polyunsaturated long chain triacylglycerols are severely reduced at temperatures above 350°C. The overlap of diacylglycerols and ceramides also remained a problem. Many of these shortcomings were eliminated by substituting a polarizable capillary column for the nonpolar capillary column. Figure 12.6

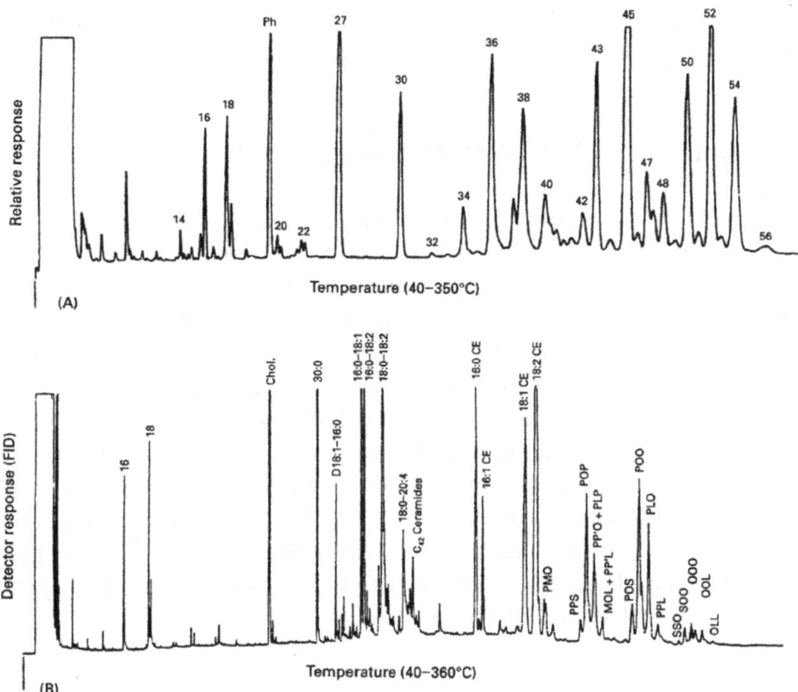

FIGURE 12.6 GLC of plasma total lipids on (A) nonpolar and (B) polarizable capillary columns. Peak identification: (A) as given in Figure 12.1; Ph, TMS ester of phthalic acid; 32, 34 and 40, TMS ethers of diacylglycerols made up of 32, 34 and 40 acyl carbons; 41 and 42, TMS ethers of ceramides with a total of 41 and 42 carbons; 43 and 47, cholesteryl esters with a total of 43 and 45 carbon atoms; 48 and 56, triacylglycerols with 48 and 56 acyl carbons. (B) as given in the figure: L, linoleic; M, myristic; O, oleic, P, palmitic, and S, stearic acids. GLC conditions: (A) Hewlett-Packard Model 5880 Gas Chromatograph equipped with dedicated capillary inlet system (fused silica needle) and a nonpolar capillary column (8 m × 0.32 mm ID flexible quartz capillary column coated with chemically bonded SE-54 methylphenylsilicone liquid phase. Carrier gas, hydrogen at 6 psi head pressure; (B) Hewlett-Packard Model 5880 Gas Chromatograph equipped with a flame ionization detector, automatic injector and fused silica capillary column (25 m × 0.25 mm ID) coated with methyl 65% phenylsilicone (OV-22). Temperature program was as described in text. Trioleoylglycerol (OOO) was eluted in 37 min.

compares the plasma total lipid profiles as obtained by GLC on (A) nonpolar and (B) polarizable capillary columns.[16] The nonpolar column yields prominent peaks for free cholesterol, diacylglycerols and ceramides, cholesteryl esters and triacylglycerols. The lipid ester classes are resolved according to the total number of carbons, very much like those demonstrated for the packed column discussed above.

Finally, two more variations in the methodology may be noted. The use of highly inactive capillary columns has permitted direct analysis of plasma neutral lipid extracts without any further derivatization.[17] Determination by gas chromatography yielded higher total cholesterol and lower triacylglycerol values than those obtained by enzymatic methods. In an application of the automated gas chromatographic method for neutral lipid carbon number profiling of marine samples, it was found necessary to include a hydrogenation step to avoid discrimination in the hydrogen flame ionization detector for highly unsaturated lipids.[18]

12.6 SUMMARY AND CONCLUSIONS

The separation and quantification of the major classes of plasma lipids by direct high temperature gas chromatography following a limited preliminary work-up is theoretically sound and practically feasible. To the biochemist, it offers an attractive possibility of multicomponent analysis at low concentration along with high resolution. To the medical investigator, it provides an opportunity for the simultaneous monitoring and study of many more metabolites than is usually possible, even with automated systems, which facilitates both diagnosis and prognosis, as well as following the progress of patients treated for a variety of plasma lipid abnormalities. To the students in the advanced biochemistry laboratory it provides immediate access to the composition, separation and quantification of plasma lipids and their alterations with diet, disease and pharmacological manipulation. It also exposes the students to the development of modern analytical methodology and its application to lipid research and clinical diagnosis.

ACKNOWLEDGMENTS

This analytical routine was developed with the collaboration of Drs. J. J. Myher and L. Marai of the Banting and Best Department of Medical Research, University of Toronto. The development was supported with funds from the Ontario Heart Foundation, Toronto; the Medical Research Council of Canada, Ottawa; and the National Heart and Lung Institute, NIH-NHLI-72-2917, Bethesda, Maryland. It was executed in the Advanced Biochemistry Laboratory of the Department of Biochemistry, University of Toronto, under the supervision of the following former biochemistry graduate students: W. C. Breckenridge, D. A. Gornall, P. Child, P. W. Connelly, F. Manganaro, S. Pind, R. Lehner and L.-Y. Yang.

REFERENCES

1. Havel, R. J. and Kane, J. P. , Introduction: Structure and metabolism of plasma lipoproteins, in *The Metabolic Basis of Inherited Disease, Volume I,* Scriver, C. F., Beaudet, A. L., Sly, W. S. and Valle, D., Eds., McGraw-Hill Information Services Company New York 1129–1164. 1989.
2. Wiebe, D. A., Lipid testing for the year 2000 and beyond, *Atherosclerosis,* 108 Suppl: S181–189. 1994.
3. Kuksis, A., Marai, L. and Gornall, D. A., Direct gas chromatographic examination of total lipid extracts, *J. Lipid Res.* 8, 352–358. 1967.
4. Kuksis, A., Myher, J. J., Marai, L. and Geher, K., Determination of plasma lipid profiles by automated gas chromatography and computerized data analysis. *J. Chromatogr. Sci.* 13, 423–430. 1975.
5. Kuksis, A., Myher, J. J., Geher, K., Hoffman, A. G., Breckenridge, W. C., Jones, G. J. and Little, J. A., Comparative determination of plasma cholesterol and triacylglycerol levels by automated gas–liquid chromatographic and Autoanalyzer methods. *J. Chromatogr.* 146, 393–412. 1978.
6. Kuksis, A., Direct gas chromatographic fractionation of mixed neutral lipids of natural origin. *Can. J. Biochem.* 42, 419–430. 1964.
7. Kuksis, A. and Myher, J. J., Gas chromatographic analysis of plasma lipids, in *Advances in Chromatography, Volume 28,* Giddings, J. C., Grushka, E. and Brown, P. R., Eds., Marcel Dekker, Inc., New York, pages 267–332. 1989.
8. Anonymous, Manual of Laboratory Operations, Lipid Research Clinics Program, Vol. 1 Lipid and Lipoprotein Analysis, National Heart and Lung Institute, National Institutes of Health, Bethesda, MD, DHEW Publication No. (NIH) 75–628, 1–81. 1975.
9. Kuksis, A., Plasma lipid profiles by automated gas chromatography, *Can. Res. & Develop.* July–August, 13–18. 1974.
10. Kuksis, A., Stachnyk, O. and Beveridge, J. M. R. Unpublished results, 1965.
11. Kuksis, A., Marai, L., Myher, J. J. and Geher, K., Identification of plant sterols in plasma and red blood cells of man and experimental animals. *Lipids* 11, 581–586.
12. Kuksis, A., Myher, J. J., Geher, K., Jones, G. J. L., Breckenridge, W. C., Feather, T., Hewitt, D. and Little, J. A. Decreased plasma phosphatidylcholine/free cholesterol ratio as an indicator of risk for ischemic vascular disease. *Arteriosclerosis* 2, 296–302. 1982.
13. Kuksis, A., Roberts, A., Thompson, J. S., Myher, J. J. and Geher, K. Plasma phosphatidylcholine/free cholesterol ratio as an indicator for atherosclerosis. *Arteriosclerosis* 3, 389–397. 1983.
14. Kuksis, A., Myher, J. J., Breckenridge, W. C. and Little, J. A. Unpublished results, 1985.
15. Marai, L., Kuksis, A. and Myher, J. J. Unpublished results. 1987.
16. Kuksis, A., Myher, J. J., and Geher, K., Quantification of plasma lipids by gas liquid chromatography on high temperature polarizable capillary columns, *J. Lipid Res.* 34, 1029–1038. 1993.
17. Lohninger, A., Preis, P., Linhart, L., Sommoggy, S. V., Landau, M. and Kaiser, E. Determination of plasma FFAs, free cholesterol, cholesteryl esters, and triacylglycerols directly from total lipid extract by capillary gas chromatography. *Anal. Biochem.* 186, 243–250. 1990.
18. Yang, Z., Parrish, C. C. and Helleur, R. J. Automated gas chromatographic method for neutral lipid carbon number profiles in marine samples. *J. Chromatogr. Sci.* 34, 556–568. 1996.

13 Analysis of Total Milk Fat

Cristina Cruz-Hernandez and John K.G. Kramer*

CONTENTS

13.1 INTRODUCTION

Milk is an oil-in-water emulsion composed of a mixture of lipids (4.1%), proteins (3.6%), carbohydrates (5%) and other minor components, i.e, vitamins and minerals.[1] Milk is also considered a colloid made of water with fat droplets mixed in it

* Corresponding author.

The lipid fraction of milk is composed mainly by triacylglycerols (TAGs), which constitute about 97.5% of milk fat. TAGs are present in the form of small globules (0.1 to 15μm) surrounded by a thin membrane composed of protein and phospholipids (PLs). Other than TAGs, milk lipids include di- and monoacylglycerols, cholesterol, glycolipids (GLs) and free fatty acids (FFAs). Milk fat contains over 400 fatty acids (FAs) that could theoretically contain over 6 million different molecular species of TAGs. The main variables among the FAs in milk fat are chain length and the number, position and configurations of the double bonds, as well as branching and the presence of other functional groups (i.e., keto or hydroxyl moieties).[2,3] Fat in milk is subject to variations in both amount and composition due mainly to nutritional conditions and factors such as species and lactation differences.[4,5] Such changes in the chemical composition can impact dairy product quality, flavor and texture.[6]

The nutritional role of milk in the human diet has traditionally been evaluated based on its overall contribution of essential and nonessential nutrients to support optimal growth and development. In recent years, milk's image as a "natural and almost perfect food" has become somehow "less perfect" due to concerns regarding its fat composition and cholesterol content. Nowadays, particular interest has focused on the identification and prevention of *trans* fatty acids (TFAs) in different foods, as these FAs have been related to increased risk of coronary heart disease (CHD).[7,8] Despite numerous studies on health concerns related to the consumption of some lipids, their health benefits are also widely investigated.

Milk retains its positive reputation, and nutritionists recognize its nutrient density and value in providing essential nutrients in a balanced diet. Cow's milk and other dairy foods are a major source of calcium and essential nutrients including potassium, phosphorous, riboflavin, vitamin B12, protein, zinc, magnesium and vitamin A.[9] Furthermore, the consumption of milk and other dairy products has been shown to help reduce the risk of chronic disease disorders including osteoporosis, hypertension, excess body weight and some cancers.[7,10,11] Conjugated linoleic acid (CLA), an FA found in dairy fat, confers a wide range of anticarcinogenic benefits in experimental animal models and is especially consistent for protection from breast cancer.[10–12] Bovine milk under normal husbandry practices may contain 2.4 to 37 mg CLA/g fat.[13] In view of the potential benefits to human health there is considerable interest in developing nutritional strategies to increase the content of such beneficial FA in milk fat as vaccenic acid (11*t*-18:1), rumenic acid (9*c*11*t*-18:1), the major CLA isomer in the fat of ruminants, and n-3 FA.[14]

A wide variety of FAs are found in milk fat; some are associated with unique health benefits, while others are commonly associated with a negative health image. To more fully understand the complex lipid composition of milk fat requires comprehensive analytical techniques to provide more detailed and correct information on the chemical composition of milk fat. This was clearly demonstrated when incorrect identification of the *trans*-16:1 isomers led to their questionable association as a risk factor for CHD in adults.[15] The same applies currently with the association of TFAs and CHD. It is still not known which TFAs are responsible for CHD, and yet no concerted effort is made to provide the detailed isomer composition necessary to properly assess this correlation.[16,17] This information could be used to explain the differences in CHD response based on the TFA isomer composition of milk fat and partially hydrogenated vegetable products.[18]

This chapter is intended to familiarize the reader with the process of extracting lipid constituents from dairy products and their subsequent detailed analysis. Milk fat analysis includes many different techniques such as lipid extraction, methylation and various forms of separation methods including thin layer chromatography (TLC), gas chromatography (GC), high performance liquid chromatography (HPLC) and argentation (silver ion, Ag^+) chromatography on TLC or HPLC phases. Performing laboratory work of this nature not only complements what is taught in undergraduate laboratories but is also a step forward in solidifying and improving learning concepts in the classroom. Exposure of the student to specialized instruments and computer-controlled software is not generally found in undergraduate teaching laboratories, but this exposure will assist students toward completing post-graduate academia and entering the work force.

13.2 THE COMPLEXITY OF MILK FAT FATTY ACID COMPOSITION

Milk fat contains more than 400 different FAs[19] and no single method is presently available to separate all of them. However, a set of comprehensive analytical techniques will be presented in this chapter that will maximize the separation of most of the FAs, particularly those associated with health benefits and possible concern. GC remains by far the most convenient method to identify and quantitate FA. However, in the past, GC separations provided only limited information on the identification and quantitation of the CLA and 18:1 isomers. The availability of highly polar capillary columns and complementary chromatographic and chemical derivatization techniques has made it possible to better evaluate milk and meat fat from ruminants.

Thorough analyses are necessary because many diet modifications have been made in rumen nutrition to improve the content of specific components without evaluation of all the resultant lipid changes in ruminant fats.[17,20,21] For example, increased concentrations of n-3 and CLA in dairy fats have been reported with the inclusion of fish meal,[22–24] fish oil[25–30] and algae.[31] Also, increased levels of CLA have been reported in milk and butter by feeding different vegetable oils.[32–38] Unfortunately, the methods used in these studies, with the exception of very few,[29,30] did not permit a complete evaluation of the CLA and 18:1 isomer composition.

A number of reports have appeared recently describing the advantages and disadvantages of different kinds of GC columns and chromatographic conditions.[15,39–42] Extensive overlap is generally observed among the different *cis*- and *trans*-18:1 isomers, as well as the many *trans* containing 18:2 and 18:3 FAs. Underestimation of the *trans*-18:1 isomers is very common because of a lack of separation of the 18:1 isomers.[43,44] A prior silver ion TLC (Ag^+-TLC) separation followed by GC analysis at reduced temperature conditions has proven to be ideal to obtain a complete analysis of the 18:1 isomers. All the *trans*-18:1 isomers were resolved except the group of 6*t*-, 7*t*- and 8*t*-18:1 using the longer GC temperature program.[21,45-48]

The introduction of the long 100 m highly polar capillary columns markedly improved the separation of the fatty acid methyl esters (FAMEs) prepared from milk fats. Between 100 and 150 different FAMEs are resolved using this type to

GC column.[21,41] Temperature programs can be effectively used to improve the separation of different FAME regions and to identify specific FAMEs. Just to cite a few examples, the coeluting 18:3 and 20:1 isomers were resolved by lowering the temperature from 180°C to 155°C,[49] the resolution of most of the *trans*-18:1 isomers was achieved by operating the GC at 120°C.[21,46-48] and identifying methyl 11-cyclohexylundecanoate and many of the *c/t*- and *tt*-18:2 isomers in the total milk fat FAMEs mixture by conduction different isothermal operations from 130 and 190°C.[42] Another important parameter in GC analysis that needs to be taken into consideration is the load applied onto the column. At low sample load the resolution of the 18:1 isomers can be improved, while at higher sample load many minor FAMEs are unequivocally identified, such as the minor 16:1, CLA and polyunsaturated fatty acids (PUFAs).[21] In the case of the CLA region, even with a very low sample load, it is not possible to separate many CLA isomers, e.g., 7*t*9*c*-, 8t10*c*- and 9*c*11*t*-CLA coelute using GC. The separation of these CLA isomers requires the use of silver ion HPLC,[50] which will be discussed in greater detail below.

The lack of separation of the two CLA isomers, 7*t*9*c*- and 9*c*11*t*-CLA, leads to a misunderstanding of rumen metabolic processes. The common intermediate in the biohydrogenation of the essential FAs, 18:2n-6 and 18:3n-3, in the rumen by rumen bacteria is 11*t*-18:1, which passes through the duodenum into the blood and is then desaturated in rumen tissues, including the mammary glands, to 9*c*11*t*-CLA by Δ9 desaturase.[51] Different clinical studies have indicated that humans can also convert 11*t*-18:1 to 9*c*11*t*-CLA.[52] However, this particular biohydrogenation process occurs predominantly when ruminants are fed high-fiber diets (pasture-fed), with minimal supplementation of concentrate. The milk fat of such cows is recognized by the high content of 11*t*-18:1, 9*c*11*t*- and 11*t*13*c*-CLA.[53,54] The addition of increased amounts of concentrates to the diet of ruminants, which generally consist of highly fermentable carbohydrate sources and vegetable oils or fish oils, alters the rumen bacterial flora to produce many *trans*-18:1 isomers other than 11*t*-18:1, specifically 10*t*-18:1.[17,55,56] Furthermore, feeding these diets will also increase CLA isomers such as 7*t*9*c*-, 9t11*c*- and 10t12*c*-CLA.[17,30,50,56] Therefore, by ignoring the levels of 7*t*9*c*- and 9*c*11*t*-CLA, one fails to recognize the symptoms of an altered rumen bacterial population that produces milk and meat fat with much reduced levels of the health-promoting FAs.[54]

Due to the complex nature of dairy fat systems and the importance of having proper analyses, it is recommended to use complementary methodologies that provide a complete analysis of total milk FAs and their isomeric composition with enough sensitivity and precision. The proposed methodology for milk fat analysis in this chapter combines such complementary techniques as GC with 100 m highly polar capillary columns, different GC temperature programs, Ag^+-TLC/GC, TLC and Ag^+-HPLC that have been proven to give reliable and complete results in this laboratory.[17,20,21,41,48]

13.3 EXTRACTION OF TOTAL MILK AND DAIRY LIPIDS

Extraction is a very common laboratory procedure used to isolate or purify a sample. Typical laboratory extractions of lipids involve the removal of substances soluble in organic solvents from an aqueous phase. The type of extractions can be solid–liquid or liquid–liquid. The distribution of a solute between two phases is an equilibrium condition described by the partition theory.[57] Liquid–liquid extractions are commonly used for milk fat extraction. In this procedure the organic substance (lipid) is soluble in an organic solvent (organic layer) and is removed from inorganic substances and the aqueous medium. An important consideration in milk fat characterization is that samples should be frozen immediately at –70°C until analyzed to avoid oxidation as well as to inhibit the action of degradative enzymes such as lipases and phospholipases that result in the release of FFA and the production of off-flavors.

When extracting lipids from milk and dairy products, it is important to use solvents that extract both neutral and polar lipids. It is also important to avoid acid digestion, which could cause isomerization of CLA. When extracting fat from solid matrices such as cheese or meat, samples should be quickly ground, preferably in the cold prior to lipid extraction, which increases the surface area of the sample to ensure complete extraction of the lipid. The preferred extraction methods are those according to Folch et al.[58] and Blight and Dyer,[59] which use different proportions of chloroform/methanol/water, i.e., 8:4:3 and 2:2:1.8, respectively. Hexane/isopropanol[60] or dichloromethane/methanol[61-63] have also been used to extract lipids from different matrices, specifically tissues.

The organic solvents used for lipid extraction should meet the following criteria. First, solvents should readily dissolve the lipids to be extracted. Second, they should not react with the lipids. Third, the solvent of choice should not react with or be miscible in water. Most solvents dissolve small amounts of water that must be removed after the extraction. Common drying agents include sodium sulfate and magnesium sulfate. Fourth, solvents should have a low boiling point to make their removal easier. It is important to minimize the researcher's exposure to organic solvents and to work in well-ventilated areas.[17]

Extraction methods should always be evaluated and compared with established methods known to give complete and quantitative extractions. TLC is often used to evaluate the qualitative extraction of the neutral and polar lipids using the developing solvents of hexane/diethyl ether/acetic acid (85:15:1) and chloroform/methanol/water (65:25:4), respectively.[21,64] Only cleaned TLC plates should be used that have been washed by developing them in chloroform/methanol (1:1). After the plates are developed, 1 cm of silica layer along the upper edge of the plate is scraped off, followed by activation at 110°C for an hour. At all times avoid greases, soaps and plastics. Preferably use only acid-washed glassware at all times.

13.3.1 Suggested Extraction

A known volume of well-homogenized milk is placed in a centrifuge tube (15 or 45 mL depending on sample size) kept in an ice bath. Add chloroform/methanol/water in a 1:2:0.8 (v/v/v) ratio; the water portion includes the water in the milk that is added. For better lipid recovery make sure the solvent-to-sample ratio is large enough (15:1 or greater), which will reduce the amount of lipid absorption onto the protein at the interphase. Mix this monophase solution well and leave standing for 2 min. After further addition of one volume of chloroform and one volume of water in that order, the solvents will separate to form a biphase system (chloroform/methanol/water; 2:2:1.8, by vol). Make sure the solution is slightly acidic, using 1 N HCl to ensure that no FFAs are present as sodium salts, and thus water soluble. The mixture is then centrifuged to clarify the lower chloroform layer.

The chloroform layer contains total lipids and the methanol/water phase, non-lipids. The chloroform layer can be easily removed using two long Pasteur pipettes inserted into one another. The total lipids are recovered by removing the chloroform using a rotary evaporator or by flushing the vessel with nitrogen when not much solvent is present. When using a rotary evaporator, a few drops of benzene can be added at the end to remove traces of water after the chloroform has been removed.[21] The total lipid content is determined by weight difference. Total lipids are dissolved in chloroform and transferred into a vial and stored at –70°C until analyzed. Do not store lipids in chloroform/methanol because this will cause transmethylation with time even at –20°C.

The rotary evaporator is equipped with a water bath that should not exceed 35°C, a dry-ice condenser, a solvent-resistant vacuum pump and another dry-ice trap to collect any organic fumes that may have escaped the initial condenser; see typical setup in Cruz-Hernandez et al.[17,21] This type of system allows for rapid removal of most of the organic solvents in minutes.

13.4 METHYLATION PROCEDURES

Intact TAGs or FFAs are generally not suitable for FA determination because they lack volatility for GC analysis. Before GC determination it is necessary to prepare volatile derivatives, and FAMEs are preferred because of their volatility and low polarity. During methylations, O-acyl and N-acyl lipids are transesterified in the presence of a catalyst and a short-chain alcohol (i.e, methanol, butanol) that replaces the glycerol or sphingosine moiety. The reaction can be carried out using an acidic (HCl, BF_3, H_2SO_4) or alkaline catalyst [$NaOCH_3$, tetramethylsilyldiazomethane (TMS-DAM), tetramethylguanidine]. Short-chain FAMEs typically present in milk are volatile, and therefore any step to reduce the solvent volume by evaporation should be avoided. The generation of FAME can be done using isolated lipids, or directly by combining extraction and transesterification in a one-step procedure, provided the samples are dry. The most commonly used catalysts are described below. The reader is referred to several good reviews on this subject.[20,21,64,65]

Sodium methoxide ($NaOCH_3$). FAMEs produced with a base (alcoholate), form an anionic intermediate, which is transformed in the presence of large excess of the alcohol (i.e., methanol, ethanol, butanol) into a new ester. FFAs, amides and alk-1-enyl ethers are not subjected to nucleophilic attack by alcohols or bases and thus are not esterified under these conditions. Milk fats contain mainly O-acyl lipids and contain CLA, but no alk-1-enyl ethers and very small amounts of amides and FFAs, and therefore, derivatizations in the presence of basic catalysts are recommended. The most useful reagent is 1 to 2 M sodium or potassium methoxide in anhydrous methanol. These solutions are stable for several months at 4°C until a white precipitate of bicarbonate salt starts to form. $NaOCH_3$ rapidly and completely converts esters to FAMEs within 10 to 15 min without isomerization of CLA. The catalyst TMS-DAM is also suitable and will methylate FFAs without causing isomerization of conjugated double bonds. However, a note of caution is appropriate, because the TMS-DAM reagent may not be pure and the resultant FAMEs will need to be purified by TLC using hexane/diethyl ether/acetic acid (85:15:1). This derivatization is not recommended when working with milk fat because of the presence of volatile FAMEs that will escape during the TLC purification step.

Hydrochloric acid (HCl). An acid catalyst is required when substantial amounts of FFA, alk-1-enyl ethers (plasmalogenic lipids) and amides (N-acyl lipids) are also present in the sample to be analyzed. Ether lipids are not dissociated from the glycerol moieties and are therefore analyzed as their isopropylidine derivatives; their presence can be detected using TLC. A solution of HCl/methanol (1 to 2 M) converts the plasmalogenic lipids (alk-1-enyl ethers) to dimethylacetals (DMA). DMAs elute just below FAME on TLC using 1,2-dichloromethane as developing solvent.[66] DMAs elute before FAME on these polar GC columns slightly ahead of the FAME with one carbon less, i.e., 18:0 DMA elutes just after the FAME 17:0.[17,21] The HCl catalyst is easily prepared and is clean, and methylations are complete within minutes except for sphingolipids (e.g., sphingomyelin) that require 1 h at 80°C. Using lower-reaction temperatures reduces the isomerization of CLA, but the methylations of all lipids may not be complete.

Acid catalyzed methylation is not recommended for milk fat analysis because it decreases the content of all the *c/t*-CLA isomers (i.e., 9*c*11*t* 18:2) produces the corresponding *tt*-CLA isomers (i.e., 9*t*11*t* 18:2) and methoxy artifacts.[67] Methoxy artifacts can also be produced from hydroxy FAs present in milk fat.[68]

Sulfuric acid (H_2SO_4). Sulfuric acid has been used as a catalyst for the preparation of isopropyl and other esters from dairy fats.[69–71] Because only the total CLA content was reported in these publications, the expected isomerization of CLA could not be assessed. The longer-chain esters have the advantage of providing flame ionization detector (FID) responses that do not require correction factors for quantitation of the FAME by GC.[69] For this reason, longer-chain alcohols have been used to prepare isopropyl and butyl esters.[72] Even though these longer-chain esters cause less interference with the solvent peaks, the resolutions of closely eluting isomers also appear to be compromised. There is also a general concern that short-chain fatty acid (SCFA) derivatives might be lost during aqueous washes if such steps are used in the derivatization methods. For this reason, nonaqueous methylation procedures are preferred.[32,73]

Boron trifluoride (BF_3). The BF_3/methanol reagent is commercially available. The catalyst is acidic and therefore suitable for the methylation of all lipid classes, including amides, FFAs and plasmalogens. However, the reagent is not stable and will result in by-products if not fresh. The resultant FAMEs should be purified by TLC prior to GC analysis. In addition, this catalyst will isomerize CLA from the *c/t*- to *tt*- forms and produce methoxy artifacts.[67] Therefore, BF_3 is not recommended for milk fat analysis.

13.4.1 Suggested Methylation

Methylation using $NaOCH_3$ (e.g., 0.5N methanolic base #33080, Supelco Inc., Bellefonte, PA) is preferred for the analyses of CLA containing lipids, as the conversion is complete within 15 min, and it does not cause isomerization of CLA. An aqueous-free system is preferred for the methylation of milk fat lipids.[32,73] Briefly, 2 mg of total milk fat is added to a 2 mL autosampler vial, the solvent is removed with a stream of N_2, and 1.7 mL of hexane and 40 μL of methyl acetate are added and mixed. Then 100 μL of $NaOCH_3$ are added. The vial should be securely capped, mixed and allowed to react for 20 min at room temperature with occasional mixing. The vial is then cooled at -20°C for 10 min after addition of 60 μL of oxalic acid (0.5 g in 15 mL diethyl ether) and thoroughly mixed. The vial is centrifuged to settle the Na-oxalate precipitate, and the upper phase is passed through a Pasteur pipette (5 3/4 in.) column containing a glass wool plug (glass wool was washed with 1:1 chloroform/methanol and dried), and a 2 cm bed of anhydrous Na_2SO_4, directly into a 2 mL autosampler vial. The FAME solution is used directly for GC analyses.

13.5 ANALYSES OF TOTAL MILK FAT BY GC

13.5.1 Suggested GC Analysis

GC methods have been widely used for total FA analysis of milk fat, and we describe here the method we use in our laboratory, which can be easily applied to different equipment. A Hewlett-Packard Model 5890 Series II gas chromatograph (Palo Alto, CA) is used with a split/splitless injection port and FID, an autosampler (Hewlett-Packard Model 7673), and a Hewlett-Packard ChemStation software system (Version A.10). We set the injection system to splitless mode and time to 0.3 min, i.e., the time the sample remains in the injection port before the carrier gas flushes it. The injector and detector temperatures were set at 250°C. The gases were H_2 as carrier gas (1 mL/min), and for the FID (40 mL/min), N_2 makeup gas (35 mL/min) and purified air (280 mL/min). A 100 m fused capillary column was used (CP-Sil 88, Varian Inc., Mississauga, ON) and the temperature program was initial temperature of 45°C and held for 4 min, programmed at 13°C/min to 175°C and held for 27 min, programmed at 4°C/min to 215°C and held for 35 min.[21,41,48] Generally, two sample loads are analyzed by GC: A 1 μL solution containing about 1 to 2 μg is injected to determine FAMEs present at low concentration, and a 3:1 dilution allows for a better resolution of the 18:1 region. The FAMEs were identified by comparison with a GC reference FAME standard (#463), spiked with a CLA mixture containing all the four positional CLA isomers (#UC-59M), and long-chain saturated FAME 21:0, 23:0, and 26:0; all FAMEs were obtained from Nu-Chek-Prep Inc. (Elysian, MN). If reference FAMEs are not available, they are synthesized, obtained from natural sources, compared with published separations or identified by GC/mass spectrometry (MS). The short-chain FAMEs are corrected for mass discrepancy using theoretical correction factors.[69]

Although complementary techniques are essential to resolve many of the FAMEs in milk fats, the use of a 100 m GC capillary column is mandatory for the analysis of the total FAME. More than 100 FAMEs can generally be identified using such a GC column and selecting conditions described here.

13.5.2 Importance of Sample Load

As mentioned before, sample load is an important factor to consider in GC analysis. At low sample load, closely eluting FAMEs will be better resolved than at high sample load, which is particularly useful in the 16:1 and 18:1 region. However, analyses of some FAMEs present at low concentrations (SCFA, PUFA, CLA) with confidence requires higher sample loads. Therefore, we often analyze milk fats at two different sample loads. There is a note of caution: milk fat FAMEs can be diluted but not concentrated because of the presence of short-chain FAMEs. Figure 13.1

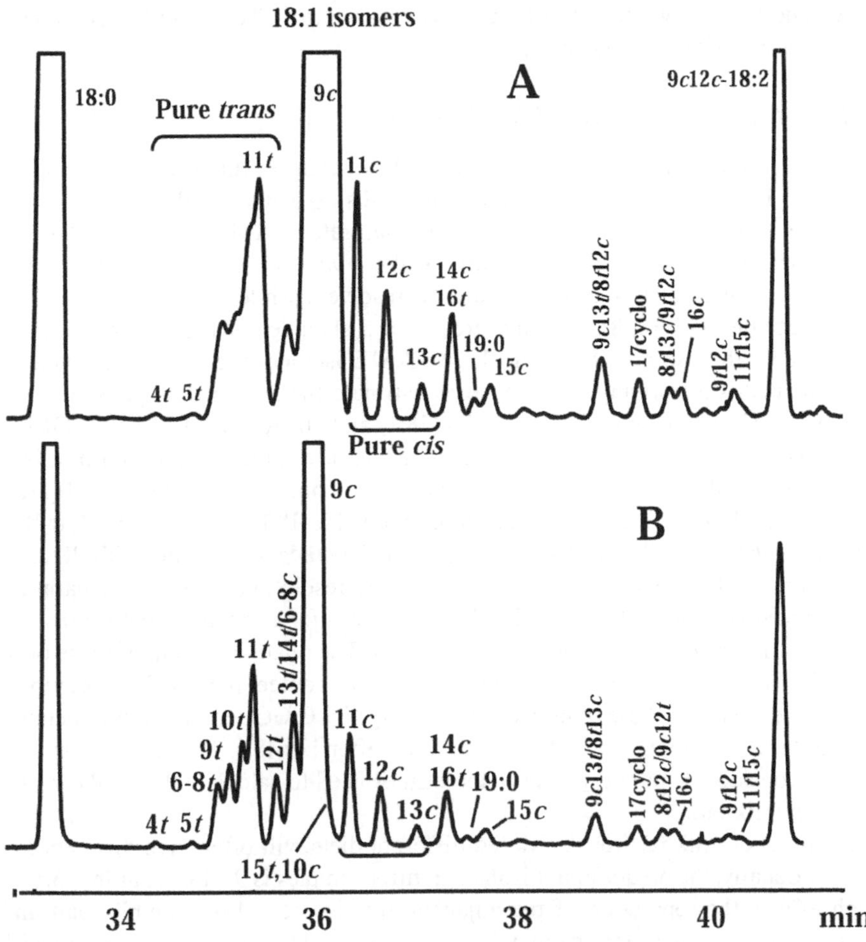

FIGURE 13.1 A partial GC chromatogram of the 18:1 FAME region using a 100 m CP Sil 88 capillary column, hydrogen as a carrier gas and a typical temperature program from 45 to 215°C. (A) High sample load of total milk fat converted to FAME; (B) low sample load of the same milk fat FAME as in (A); a baseline resolution is observed between 11t- and 12*t*-18:1.

shows chromatograms of the 18:1 FAME region of milk fat at high (Figure 13.1A) and low (Figure 13.1B) sample load. The separation of the 13-14*t*/6-8*c* peak is possible at low, but may not be resolved at high sample load because of the relative high concentration of the adjacent methyl oleate (9*c*-18:1) peak.

13.5.3 Importance of Temperature Program

The relative elution order of FAMEs changes with column temperature. This has been used effectively to resolve different groups of FAMEs. However, such manipulations have not been able to resolve all the *cis*- and *trans*-18:1 isomers. For instance, at isothermal condition of 170 or 175°C, 13*t*/14*t*-18:1 elutes just before and 15*t*-18:1 just after the major 9*c*-18:1 peak.[42,74] On the other hand, at 150°C, 13*t*/14*t*-18:1 elutes with 9*c*-18:1 and 15*t*-18:1 with 11*c*-18:1, and at 130°C, 13*t*/14*t*-18:1 elutes between 9*c*-18:1 and 11*c*-18:1, while 15*t*-18:1 elutes with 12*c*-18:1.[42] The separation of the 18:3 and 20:1 isomers[49] was cited above.

13.5.4 Analysis of the 18:1 FAME Region

Typical separation of the FAMEs in the 18:0 to 18:2n-6 region is shown in Figure 13.1B. This region contains overlapping peaks associated with *cis*-18:1, *trans*-18:1, *tt*-18:2 and *c*/*t*-18:2 isomers, and some saturated FAMEs such as 19:0 and 11-cyclohexylundecanoic acid (17 cyclo). Of the *trans*-18:1 isomers, 4*t*-, 5*t*-, 6-8*t*-, 9*t*-, 10*t*-, 11*t*- and 12*t*-18:1 can generally be resolved, while the 13*t*- and 14*t*-18:1 isomers coelute with 6*c*-8*c*-18:1 and are often a shoulder of the leading edge of the major 9*c*-18:1 isomer. As such, these FAME isomers are poorly resolved or overestimated when attempts are made to estimate the area of this peak. On the other hand, 15*t*-18:1 coelutes with 9*c*-/10*c*-18:1, and 16*t*-18:1 with 14*c*-18:1. Of the *cis*-18:1 isomers, 9*c*-18:1 is the most abundant, but it coelutes with two minor isomers, 15*t*- and 10*c*-18:1. Some of the other *cis*-18:1 isomers are well resolved (11*c*-, 12*c*- and 13*c*-18:1), while the remaining coelute, 14*c*-18:1 with 16*t*-18:1, 15*c*-18:1 with 19:0, 16*c*-18:1 with 9*c*12*t*-18:2, and methyl octadec-17-enoate with 18:2n-6 (see Figure 13.1B). There is generally a baseline resolution (complete separation between peaks) between 11*t*- and 12*t*-18:1 that allows for an accurate determination of some of the *trans*-18:1 isomers (4*t*- to 11*t*-18:1). The relative content of these pure *trans*-18:1 isomers is used to calculate the total *trans* content in milk fat by comparing these results with those obtained after the Ag^+-TLC/GC separation (see below). Another base resolution is seen before 11*c*- and after 13*c*-18:1. The region from 11*c* to 13*c* is considered "pure *cis*" and is sometimes used to calculate the total *cis* 18:1 content in milk fat.

Analysis of milk fat from cows fed different diets will often give different FA profiles. Usually, the predominant isomer in milk samples is 9*c*-18:1 (Figure 13.1B), which affects the separation of the adjacent 18:1 isomers. Low sample load and different temperature programs may help to improve the separation of certain 18:1 isomers. The feeding of high-concentrate diets rich in PUFAs will increase the 10*t*-18:1 isomer, while pasture feeding will drastically increase 11*t*-18:1. In both cases, a clear resolution of these two isomers (10*t*- and 11*t*-18:1) is not possible unless a prior separation using Ag^+-TLC is performed.[20,21]

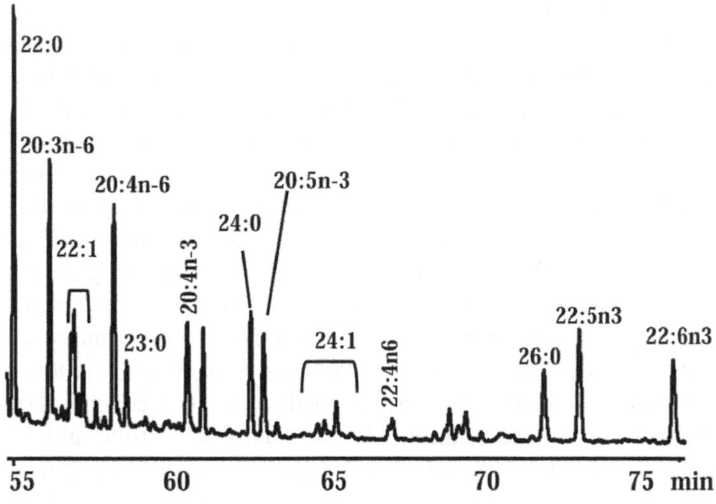

FIGURE 13.2 A partial GC chromatogram of the 22:0 to 22:6n-3 region of total milk FAME from cows fed fish meal using a 100 m CP Sil 88 capillary column, hydrogen as a carrier gas and a typical temperature program from 45 to 215°C.

13.5.5 Analysis of the PUFA Region

Analysis of the PUFAs in milk fats is possible by GC but it may require higher sample loads because of their low content. Figure 13.2 shows the 22:0 to 22:6n-3 FAME region. When working with a low sample load, some of these PUFAs may be overlooked as baseline noise, or not identified with confidence. The addition of fish oil or fish meal in many studies is designed to increase the level of long-chain PUFAs, particularly DHA (22:6n-3) and EPA (20:5n-3). The transfer efficiency of these PUFAs into milk fat is not high[75] and levels of only 0.2 to 0.4% are common. But an accurate determination of these PUFA, even at low levels, is necessary to evaluate the contribution of these health-promoting FAs.

13.6 AG^+-TLC Separation of Geometric Isomers

Argentation chromatography is based on the reversible formation of a weak charge-transfer complex between a silver ion and a double bond. A sigma bond is formed between the silver ion and the pi electrons of double bonds; *cis* forms a stronger bond than *trans* bonds. The strength of the complex depends on the accessibility of the electrons in the orbitals as well as on steric hindrance. This enables the fractionation of lipid molecules according to the number and geometry of the double bonds in the FAMEs; the position of the double bond has less of an effect.[76,77] Silver ion-impregnated TLC plates are easily prepared and have the advantage of separating FAMEs based on the number of isolated double bonds regardless of chain length and the geometry of such double bonds (*cis* and trans).[77,78] In general, the migration of FAMEs with two double bonds depends on the *cis* bonds. For example, a *tt*-18:2 isomer will migrate with the mono *trans* FAMEs, a *c/t*-18:2 with mono *cis* FAMEs, and a c/c-18:2 with methyl linoleate. The choice of developing solvent depends on the separation required. The separation of saturated and monounsaturated *trans* and *cis* FAMEs can be achieved using lower proportions of hexane and diethyl ether (90:10), while higher levels of diethyl ether (70:30) are necessary to resolve the PUFAs.

There are several reasons that a prior separation of the mono *trans* FAMEs is highly recommended. In the first place, there is extensive overlap of 13*t*- to 16*t*-18:1 isomers with the *cis*-18:1 isomers that will lead to underestimation of up to 25 to 30% if they are ignored.[44] Second, the *trans*-18:1 isomers of 6*t*-11*t*-18:1 are not well resolved, particularly when either 11*t*- or 10*t*-18:1 is relatively high compared with the other isomers.[17,21] Reducing the sample load or using different temperature programs will only partly improve the identification. For complete and reliable results, a combination of Ag^+-TLC and GC should be used to resolve and identify all the *trans*- and *cis*-18:1 isomers independently. Although less frequently used, separations can also be performed using Ag^+-HPLC columns.[79,80] Total milk FAMEs can be fractionated into saturates, mono-*trans* and mono-*cis*- FAMEs plus CLA using Ag^+-TLC.[21,48,70,74]

13.6.1 Suggested Ag^+-TLC Analysis

Commercial precoated silica G plates (20 × 20 cm × 0.25 mm thickness; Fisher Scientific, Ottawa, ON) are prewashed with chloroform/methanol (1:1, v/v), and then activated at 110°C for 1 h. All operations with Ag^+-TLC should be conducted under subdued light conditions. The plates are soaked in a 5% silver nitrate solution in acetonitrile (w/v) for 20 min, then air dried and activated at 110°C for 1 h prior to use. Mixtures of total FAMEs are applied onto the TLC plates (0.4 mg/cm) using a TLC streaker (Alltech Corporation, Deerfield, IL). A thin glass capillary or a Pasteur pipette can also be used to apply the sample across the plate, although the application should be in as straight and narrow a band as possible to maximize the resolution of the different FAMEs. TLC plates are immersed into the developing solvent hexane/diethyl ether (90:10, v/v). Bands are visualized after development by spraying the plate with a dilute solution of 2′,7′-dichlorofluorescein in methanol (w/v), dried under a stream of N_2, to avoid lipid oxidation and examined under UV light (234 nm). Figure 13.3 shows a typical Ag^+-TLC separation of total cheese fat FAMEs. The bands corresponding to saturated FAMEs (top of plate), *trans*-MUFA and *cis*-MUFA plus CLA are identified, scraped off and collected in separate Pasteur pipette (5 3/4 inch) columns containing a glass wool plug. The silica gel is eluted with chloroform, chloroform/methanol (1:1) and finally methanol. The solvent is removed under a stream of N_2. To remove the dye and dissociate any remaining Ag^+/PUFA complexes, lipids are partitioned in chloroform/methanol/1 N aqueous HCl solution (1:1:0.9). The chloroform phase is removed and dried using a stream of N_2. Samples are dissolved in hexane and analyzed by GC using a stepwise temperature program.[48] Initial temperature was 120°C held for 200 min, then increased to 150°C at 15°C/min and held for 70 min, again increased to 175°C at 15°C/min and held for 60 min, and finally increased to 220°C and held for 50 min. The isolated *cis* and *trans* fractions are analyzed at low and high sample loads to resolve the major and minor FAME isomers, respectively. The isolated *cis* and *trans* fractions by Ag^+-

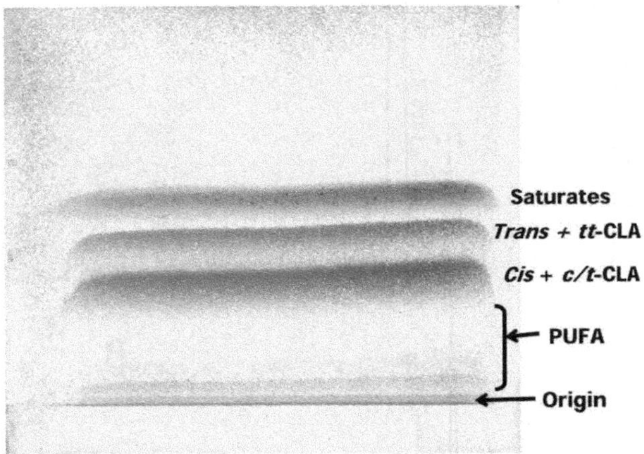

FIGURE 13.3 A typical Ag^+-TLC separation of the total milk fat FAMEs using the developing solvent hexane/diethyl ether (90:10). After development the bands were visualized by spraying the TLC plate with H_2SO_4/methanol and charring it.

TLC can also serve as GC standards to identify and establish the *trans* and *cis*-18:1 and 16:1 isomers in total milk fat separations.

13.6.2 Analysis of the *Trans*-18:1 Isomers

The isolated *trans* fraction by Ag^+-TLC was analyzed by GC using the 100 m highly polar capillary column and the temperature program from 45 to 215°C (Figure 13.4B) and compared with the total milk fat FAME analyzed under the same conditions (Figure 13.4A). By comparing the two GC separations (total milk fat FAMEs and Ag^+-TLC fractions) it is now possible to clearly recognize the extent of the overlap of the *cis*- and *trans*-18:1 isomers and identify with certainty their elution times (Figure 13.5). Furthermore, it is clearly evident that the resolution of 6*t*-8*t*- to 11*t*-18:1 is only partial, and that the isomers 6

operated at 120°C, as demonstrated for *trans*-16:1 below. For analysis of the *trans*-20:1 isomers see Precht and Molkentin.[82] Without a prior Ag^+-TLC separation the *trans*-18:1 isomers were often misidentified. For example, in a number of publications, 6*t*-18:1 was reported,[22,26,28,81] even though this isomer cannot be resolved by GC.

t-8*t*-18:1 are not resolved at all by GC under any condition, while 13*t*-/14*t*-18:1 can be separated, but only at 120°C (Figure 13.5).[15,21,46,48] The lack of separation of 6*t*-8*t*-18:1 has remained a challenge. Precht and Molkentin[47] suggested that the lack of separation of the 6*t*-8*t*-18:1 was due to the relatively small content of the 6t and 7t isomers. Using the combination of Ag^+-TLC and GC made it possible to resolve not only 13*t*- and 14*t*-18:1, but also 10*t*- and 11*t*-18:1, which in many milk fats is difficult to identify and quantitate because of lack of clear separation of these two isomers (Figure 13.5A and Figure 13.5B). The analysis of the *trans* isomers of 16:1, 20:1 and 22:1 requires the application of higher sample loads onto the GC

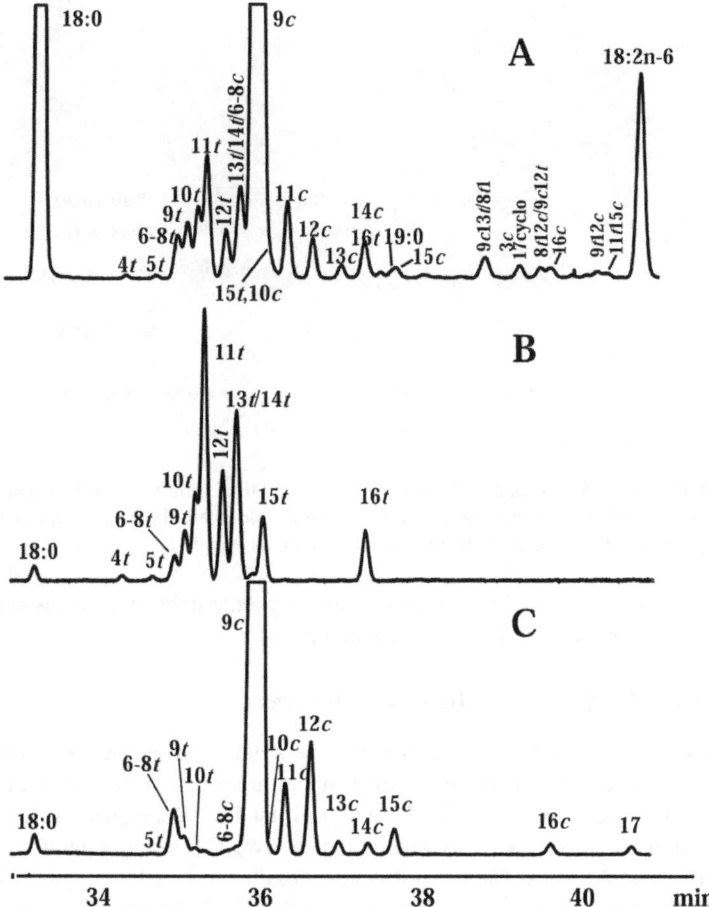

FIGURE 13.4 A partial GC chromatogram of the 18:1 FAME region of total milk fat FAMEs (A) and the *trans* (B) and *cis* (C) fraction isolated by Ag^+-TLC. The same capillary column (100 m CP Sil 88) and GC conditions were used for each of the analyses. The overlap of the *trans*- and *cis*-18:1 isomers is evident when milk fat FAMEs are analyzed by GC.

13.6.3 Analysis of the *Cis*-18:1 Isomers

A typical separation of the *cis*-18:1 isomers from the *cis* fraction isolated by Ag^+-TLC is shown in Figure 13.4C. All the *cis*-18:1 isomers are resolved using the temperature program from 45 to 215°C, except 10*c*- and 6*c*-8*c*-18:1. Using a low temperature program starting at 120°C, better resolutions of all the *cis*-18:1 isomers are obtained, but even then 6*c*-8*c*-18:1 remains unresolved and often 10*c*-18:1 remains as an unresolved peak on the slope of 9*c*-18:1 because of the large difference in their relative amounts. There is evidence that several of the minor *tt*- and *c*/*t*-18:2 isomers in milk fat elute starting with 13*c*-18:1.[42,67]

It should be noted that the separation of the bands using Ag^+-TLC might not be complete. For example, small amounts of *trans*-18:1 were evident in the *cis* band

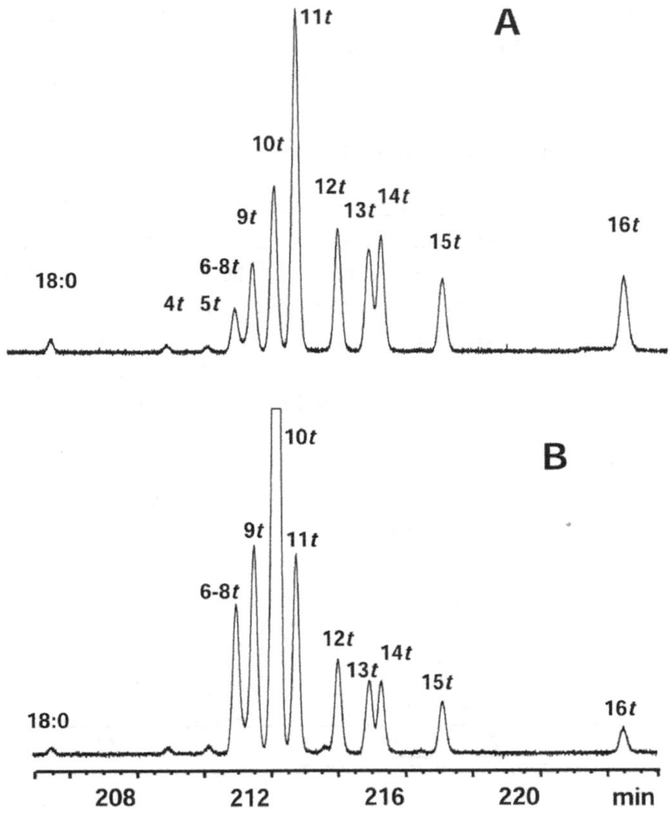

FIGURE 13.5 A partial GC chromatogram of the *trans*-18:1 isomers from the *trans* fraction isolated from two milk fats by Ag^+-TLC that were separated using a stepwise GC program starting at 120°C. (A) Milk fat from cows fed a control diet. (B) Milk fat obtained from cows fed a fish meal diet.[21]

(Figure 13.4C). Initially, it was suspected this was due to accidental inclusion of some of the *trans* band during isolation by Ag^+-TLC. However, these *trans*-18:1 isomers could not be removed even after two repurifications of the *cis* band.[21] An alternate explanation is a partial separation of some *trans*-18:1 isomers on the Ag^+-TLC plates in which the *trans*-18:1 isomers 5*t*- to 9*t*-18:1 elute near the bottom of the *trans* band or are migrating with the *cis* isomers. This partial separation of some *trans*-18:1 isomers into the mono *trans* and mono *cis* fractions on Ag^+-TLC will have quantitative implications because it would underestimate the *trans*-18:1 isomers migrating with the *cis* FAMEs.

An additional observation is noteworthy when analyzing the *trans* and *cis* fractions isolated by Ag^+-TLC separation and analyzed by GC. They all contained trace amounts of 18:0. The origin of 18:0 is not clear whether it is residual 18:0 on the Ag^+-TLC plates from the saturated band that migrate past that area, or is formed in the GC injection port.[21] Hydrogenation was suspected as heat and H_2 (carrier gas) are present in the injector port, despite the fact that the inlet insert and the injector are protected with a silylated coating. Whatever the cause, the very small amount of 18:0

present in all unsaturated FAME fractions was rather useful because it served as an internal reference to compare the GC separations. This explanation is supported by our findings that 18:0 is virtually absent when He was used as carrier gas.[41]

13.6.4 Analysis of the 16:1 FAME Region

In Figure 13.6A the 16:0 to 18:0 region of a typical milk fat is obtained using the temperature program described above (45°C to 215°C). In the 16:1 region there is an overlap of the iso- and anteiso-17:0 saturates and the *cis*- and *trans*-16:1 isomers that can only be clarified using the Ag^+-TLC isolated fraction. The GC separations at 120°C of the *cis* and *trans* fractions isolated by Ag^+-TLC are shown in Figure 13.6B and Figure 13.6C, respectively. Similar results were obtained from human milk samples.[15] The relative elution of the *cis*- and *trans*-16:1 isomers is slightly different

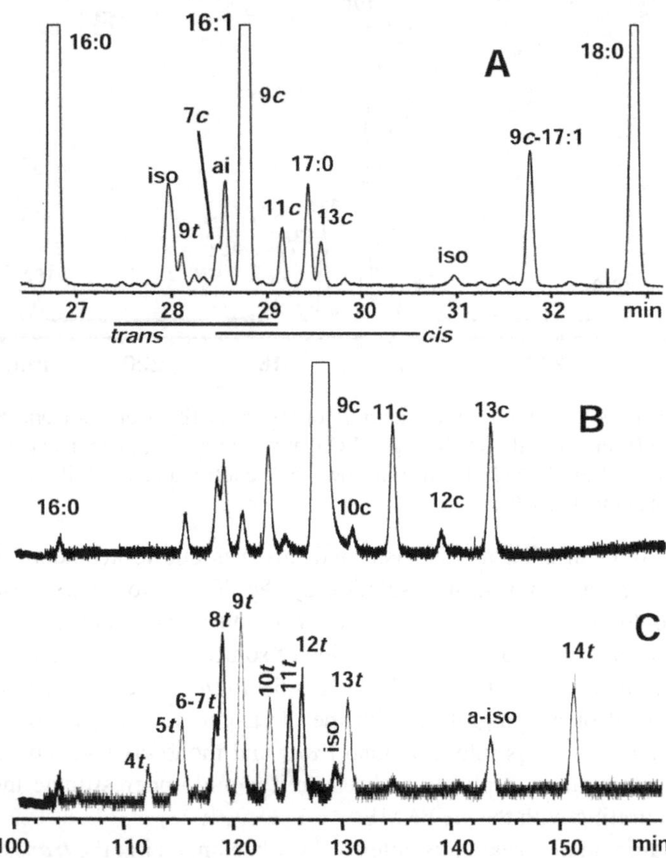

FIGURE 13.6 A partial GC chromatogram of the 16:0 to 18:0 FAME region of total milk FAME from cows fed a control diet using a 100 m CP Sil 88 capillary column, hydrogen as a carrier gas and a typical temperature program from 45 to 215°C (A). (B) and (C) are partial GC chromatograms of the 16:1 region of the cis and *trans* fractions, respectively, which were isolated from total milk fat FAME using Ag^+-TLC. A stepwise GC program was used for both the cis and *trans* fractions starting at 120°C.

in the temperature program (Figure 13.6A) compared with the 120°C conditions (Figure 13.6B and Figure 13.6C). The content of the 16:1 isomers in the *cis* and *trans* fractions isolated by Ag^+-TLC are about 1/30th of the content of the *trans*-18:1 isomers and hence the *trans* fractions need to be concentrated appropriately for GC analysis.

13.6.5 Calculation of Monounsaturated FAME

The total *trans*-18:1 content of any milk fat was calculated as follows. First, the GC separation of the total milk fat FAMEs was evaluated (Figure 13.1 or Figure 13.4). The sum of the relative concentration of 4*t*- to 11*t*-18:1 was assigned the value X%. Next, the total milk fat FAMEs were separated by Ag^+-TLC and analyzed by GC at 120°C (Figure 13.5), and all the *trans*-18:1 isomers were integrated. The area of 4*t*- to 11*t*-18:1 is equivalent to X% because it represents the same isomers from the milk fat. Therefore, the total *trans*-18:1 content can be calculated by multiplying X% (obtained from the analysis of total milk fat FAME) by the ratio of the two areas of (4*t*- to 16*t*-18:1) and (4*t*- to 11*t*-18:1) as follows:

Total *trans*-18:1 = (X%) (4*t*- to 16*t*-18:1) / (4*t*- to 11*t*-18:1)

To calculate the *trans* MUFA other than 18:1, the sum of 15*t*- and 16*t*-18:1 can be used as internal reference for comparison, because these two *trans*-18:1 isomers are well resolved at low (Figure 13.5) and high sample load (Figure 13 in reference 21) operated at 120°C. Knowing the relative concentration of 15*t*- plus 16*t*-18:1 allows one to determine the relative content of the *trans*-16:1, *trans*-20:1, and *trans*-22:1 isomers.

13.7 CLA Isomer Analysis by Combination of GC and Ag^+-HPLC

CLA is a collective term used to describe one or more of the C18 dienoic positional and geometric FA isomers having a conjugated double bond system, i.e., a single bond separating two double bonds (-CH=CH-CH=CH-). Milk fat contains numerous *c,t*-, *t,c*-, *cc*- and *tt*-CLA isomers at positions 7,9-, 8,10-, 9,11-, 10,12-, 11, 13- and 12,14-18:2.[83–85] The 9*c*11*t*-CLA isomer is the predominant natural CLA isomer accounting for about 65 to 90% of the total CLA in the milk and meat fat of ruminants.[17,50] This CLA isomer has also been associated with anticarcinogenic properties.[10,12]

Silver ion HPLC (Ag^+-HPLC) utilizes columns packed with 5–10 μm Nucleosil SA™ (phenylsulfonic acid groups bonded to a silica substrate) or similar substrates in which the sulfonic acid protons have been exchanged with silver ions (Ag^+). A nonpolar mobile phase is used to prevent the elution of silver ions. This column represents a useful technique for the separation and isolation of *cis* and *trans* geometric and positional isomers of FAMEs and TAGs[86] and has been successfully used for the separation of all the CLA isomers.[50,84,87]

A combination of GC and Ag^+-HPLC analyses is mandatory to resolve and identify all the CLA isomers. A brief description of the Ag^+-HPLC method follows. The mixture of four positional CLA isomers is used as the standard to evaluate the resolution of all the CLA isomers by Ag^+-HPLC (#UC-59M from Nu-Chek-Prep Inc., Elysian, MN). For identification of the FAMEs by GC, a reference FAME mixture (#463) and the FAME 21:0, 23:0 and 26:0 were prepared (all FAMEs are available from Nu-Chek Prep).[41] The elution order of the CLA isomers by GC are *c/t*<*cc*<*tt* and by Ag^+-HPLC are *tt*<*c/t*<*cc*. The identification of the individual CLA isomers was established by isolation of the individual isolated CLA isomers and characterization by GC/MS and GC/FTIR,[83,87] by isomerization of the CLA isomers using iodine[84,85] and by chemical synthesis.[88,89] To investigate the minor CLA isomers by GC, a higher sample load of total FAME should be used, considering that the total content of all CLA isomers generally comprise only 0.5% of total FAMEs. Under special feeding condition it may approach 4% of total milk fat.[37]

13.7.1 Suggested Analysis

The CLA isomers are separated using an HPLC system (Model 1100, Agilent Technologies, Palo Alto, CA) equipped with a quaternary pump (G1311A), autosampler (G1313A), diode array detector (G1315A) and Hewlett-Packard ChemStation software system (Version A.10). Three ChromSpher 5 Lipids (4.6 mm i.d. × 250 mm stainless steel; 5 μ particle size) analytical silver-ion impregnated columns (Varian Inc., Mississauga, ON) are connected in series to improve the resolutions of the isomers. The mobile phase is hexane containing 0.1% acetonitrile and 0.5% diethyl ether (or tertiary butyl methyl ether), and is operated isocratically at a flow rate of 1.0 mL/min. The mobile phase is thoroughly mixed at the beginning and thereafter is stirred continuously using a magnetic stirrer. The diode array detector is operated at a wavelength of 233 nm for the identification of conjugated FAMEs, at 205 nm for unsaturated (nonconjugated) FAMEs, and at 268 nm for conjugated trienoic FAs. The elution order of the CLA isomers was described previously.[84,85,87-89] The columns should be cleaned daily by flushing them with 1% acetonitrile in hexane for 1 h, followed by return to the mobile phase of 0.1% acetonitrile for 1 h before using them for the next analyses.

13.7.2 Determination of All CLA Isomers

Figure 13.7 shows the importance of the GC and Ag^+-HPLC as complementary analytical techniques. By GC the 9*c*11*t* is coeluting with the 7*t*9*c* and 8*t*10*c* isomers, but the 9*t*11*c* isomer is well resolved (Figure 13.7A). By Ag^+-HPLC it is possible to separate 7*t*9*c* from 9*c*11*t*, but now 9*c*11*t* is coeluting with the 9t11c (Figure 13.7B), hence the importance of using both methods for the analysis of these CLA isomers. These results clearly indicate that there are three GC peaks in the CLA region (Figure 13.7A) that are composed of unresolved CLA isomers: the so called "9*c*11*t*-CLA" peak that may contain 7*t*9*c*-, 9*c*11*t*- and 8*t*,10*c*-CLA; the "9*c*11*c*-CLA" peak that also contains 11*t*13*c*-CLA; and the combined *tt*-CLA peak that contains all the *tt*-CLA isomers from 7,9- to 10,12-CLA. All the *tt*-CLA isomers, both geometric isomers of 12,14- and 11,13-CLA, and 7*t*9*c*-CLA, are well resolved using Ag^+-HPLC. On

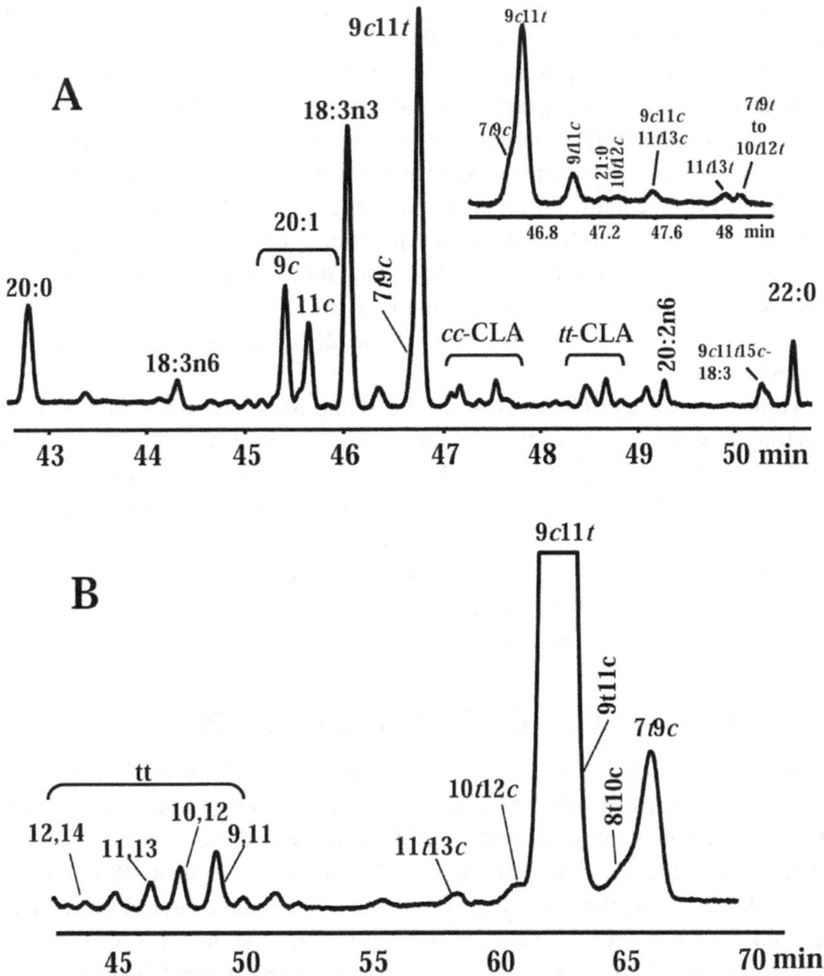

FIGURE 13.7 A partial GC chromatogram of the 20:0 to 22:0 FAME region of total milk fat FAMEs using a 100 m CP Sil 88 capillary column, hydrogen as a carrier gas and a typical temperature program from 45 to 215°C. The insert shows an enlargement of the CLA region. The lower graph shows a partial Ag^+-HPLC separation of the *cis/trans* and *trans/trans* CLA region using three Ag^+-HPLC columns in series and UV detection at 233 nm.

the other hand, there are several CLA isomers that are well resolved by GC, such as 10*c*12*t*-, 9*t*11*c*-, 10*t*12*c*-, 10*c*12*c*-, 11*c*13*c*-, 12*t*14*t*- and 11*t*13*t*-CLA (Figure 13.7A).

With the information of the GC and Ag^+-HPLC separations it is possible to calculate the concentration of each CLA isomer. The GC results are used to quantitate the CLA content, while the Ag^+-HPLC results are used to provide the relative abundance of most of the individual CLA isomers, and some of the isomers require the combination of techniques to determine their value such as 9*t*11*c*- and 7*t*9*c*-18:2.[21] A proper analysis of each CLA isomer is mandatory to avoid overestimation of

9c11*t*-CLA and provide the concentration of each of the other CLA isomers considering that each of the CLA isomers differs in its biological response.

The presence of 21:0 poses a challenge because this FAME elutes in the CLA region of the GC chromatogram. The concentration of 21:0 in natural dairy fats is generally similar to the concentration of the minor CLA isomers among which it elutes. 21:0 elutes anywhere between 9*t*11*c*- and 10*c*12*c*-CLA depending on the GC column and the temperature program used.[17,20,40,67,82,90] Therefore, the addition of 21:0 FAME to the mixture of the four positional CLA isomers is highly recommended. We have custom-made a GC reference mixture in our laboratory that contains the GC FAME standard (#463), the mixture of four positional CLA isomers (#UC-59M), and the FAMEs of 21:0, 23:0 and 26:0; all lipids were obtained from Nu-Chek Prep.

Although the major CLA isomer in dairy and beef fats is generally 9*c*11*t*-18:2, reporting only the content of this isomer based on a single GC separation overestimates this isomer by 10 to 15% of total CLA, because the contributions of 7*t*9*c*- and 8*t*10*c*-18:2 are included.[50,91] Furthermore, the importance of the other CLA isomers is being recognized in nutrition and health, and therefore these need to be determined. Due to the coelution of the 7*t*9*c*- and 9*c*11*t*-18:2 isomers and the lack of separation of several other CLA isomers by GC, Ag^+-HPLC has become a mandatory technique in the analysis and resolution of all the CLA isomers.[17,21,48,50,91]

13.8 TYPICAL RESULTS OF COMMERCIAL CHEESE

Using the combination of analytical techniques described above, the FA composition of four commercial cheese samples was determined (Table 13.1). The results indicate that the content of total saturated FAs of cheese lipids was 65%, but it contained substantial amounts of SCFA (≤14:0, 21.9%). The total *trans*-18:1 content was 3.9% with 11*t*-18:1 as the major and 10*t*-18:1 as the second most abundant isomers. The total CLA content was 0.54% of total milk fat, of which 70% was 9*c*11*t*-18:2. Several other CLA isomers were present (expressed as percentage of total CLA), such as 7*t*9*c*- (8.4%), 9t11c- (2.8%), 10*t*12*c*- (1.9%), and *tt*-CLA (16.8%).[17,54] The total PUFA content of 3.5% was mainly composed of n-6 PUFA (2.2%) with minor amounts of n-3 PUFA (0.26%), and the n-6 to n-3 ratio was about 5.

13.9 ANALYSIS OF PHOSPHOLIPIDS IN DAIRY SAMPLES

In the analysis of milk fats the phospholipids are generally ignored because they are present only at low levels, i.e., at 0.2 to 1% of total milk fat, and they are generally associated with the membrane components of the fat globules. Their FA composition is uniquely different from that of milk fat TAGs. The term PL denotes any complex lipid that on hydrolysis yields phosphoric acid. PLs can be divided into two classes, depending on whether they contain a glycerol or a sphingosyl backbone.[92] Upon hydrolysis, both PL groups yield FA, an organic base or a polyhydroxy compound. There are three major PLs, phosphatidylcholine (PC), phosphatidylethanolamine (PE) and sphingomyelin (SM), and two minor PL subclasses in milk fat,

TABLE 13.1
Fatty Acid Composition (Relative %) of Commercial Cheese

SFAs	Rel %	MUFAs	Rel %	PUFAs	Rel %
4:0	2.94	12:1	0.06	9*c*13*t*/8*t*12*c*-18:2	0.16
5:0	0.02	9*c*-14:1	1.31	8*t*13*c*/9*c*12*t*-18:2	0.10
6:0	1.51	9*t*-16:1	0.07	9*t*12*c*-18:2	
7:0	0.04	7*c*-16:1	0.18	11*t*15*c*-18:2	0.06
8:0	1.26	9*c*-16:1	1.58	18:2n-6	1.98
9:0	0.05	9*c*-17:1	0.26	18:3n-6	0.03
10:0	2.60	4*t*-16*t*-18:1	3.93	18:3n-3	0.3
11:0	0.33	4*t*-	0.03	Total CLA	0.54
12:0	2.96	5*t*-	0.04	9*c*11*t*-	0.38
13:0iso	0.03	6*t*-8*t*-	0.15	7*t*9*c*-	0.05
13:0ai	0.09	9*t*-	0.32	9*t*11*c*-	0.02
13:0	0.13	10*t*-	0.57	10*t*12*c*-	0.01
14:0iso	0.11	11*t*-	1.15	7*t*9*t*-11*t*13*t*-	0.09
14:0	9.89	12*t*-	0.41	20:3n-9	0.02
15:0iso	0.19	13*t*-	0.36	20:3n-6	0.08
15:0ai	0.39	14*t*-	0.40	20:3n-3	0.03
15:0	1.12	15*t*-	0.26	20:4n-6	0.10
16:0iso	0.26	16*t*-	0.27	20:5n-3	0.04
16:0	28.37	9*c*/10*c*-18:1	20.32	22:4n-6	0.02
17:0iso	0.33	11*c*-18:1	0.69	22:5n-6	0.02
17:0ai	0.46	12*c*-18:1	0.32	22:5n-3	0.06
17:0	0.57	13*c*-18:1	0.13	22:6n-3	0.03
18:0iso	0.07	14*c*-18:1	0.06	Total PUFAs	3.54
18:0	11.00	15*c*-18:1	0.18	n-6 PUFA	2.22
19:0	0.30	16*c*-18:1	0.14	n-3 PUFA	0.46
20:0	0.12	17-18:1	0.10		
21:0	0.03	9*c*-20:1	0.09		
22:0	0.06	11*c*-20:1	0.07		
23:0	0.02	11*c*-22:1	0.01		
24:0	0.03	13*c*-22:1	0.03		
Total SFAs	65.21	15*c*-24:1	0.01		
		Total MUFAs	29.48		

phosphatidylinositol (PI) and phosphatidylserine (PS). Each of these PLs contains about 25 to 30 different FAs.[1]

The methodologies necessary to accurately measure the small amounts of PLs in milk represent a unique challenge to lipid analysts because of the great relative difference in amounts between TAGs from milk fat (98%) and the PLs (1%) associated with the membrane lipids. Several extraction and purification methods of the individual PL classes from milk will be presented and compared here.[93–97]

Isolation of the PL from milk fat has been accomplished using preparative TLC, adsorbent chromatography on silicic acid, florisil, alumina, partition chromatography and HPLC methodologies.[98] Each of these methods is associated with its own limitations. Preparative TLC is sensitive to sample overload and is time consuming. Traditional column chromatography is time consuming and requires large amounts of solvents. HPLC has the downside of requiring expensive equipment and often requires more than one column to provide a complete separation of all lipid classes.[99,100]

Micro-preparative applications on TLC have been widely used as an easy way to separate different lipid classes. The availability of commercial precoated plates yields reproducible separations of the complex lipids. Commercially available silica gel H TLC plates need to be used to prevent streaking of PS because it contains the zwitterion. Two- or three-dimensional TLC procedures can be used to resolve all the PLs.[98,101] The two-dimensional TLC separations generally give better resolution than a single HPLC run.[98,101]

Solid phase extraction (SPE) is another common method used to separate lipids. SPE columns are made with impermeable plastic material and packed with a variety of adsorbents under trade names such as Bond Elut™ (Analytichem International, Harbor City, CA), Sep-Pak™ (Waters Associates, Milford, MA) and Supelclean ™ (Supelco Inc., Bellefonte, PA).[102] The commercially prepacked columns have uniform reproducible properties that make them useful for many applications. By using appropriate columns, under proper experimental conditions it is possible to isolate and concentrate the lipids of interest by a combination of solvents.[96,99] It is usually easier from a technical standpoint to obtain single lipid classes in a pure state after a preliminary fractionation into neutral lipids, PL or glycolipids using column chromatography with silica gel as adsorbent.[99,103] Due to the small amounts of PL in milk fat it is important to first concentrate the PL fraction before separating the individual PL classes.

13.9.1 Separation of PL Using Three-Directional TLC

Three-directional TLC has proven to be a rapid and effective method to separate all the different lipids on a single silica gel H plate.[101] The three-directional TLC method described below was applied to the separation of all the neutral lipids and PLs in milk fat.

13.9.1.1 Suggested TLC Method

Commercial precoated silica H plates (20 cm × 20 cm × 0.25mm thickness; Analtech Scientific, Newark, DE), are prewashed with a chloroform/methanol (1:1, v/v) mixture and activated at 110°C for 1 h prior to use. Silica gel H plates without binder are used instead of silica gel G plates with binder (sodium sulphate) because the acidic PL will cause streaking. Lipids are extracted as previously described. Total milk lipids (about 10 mg) are applied onto the lower left-hand corner of the plate, 2 cm from both edges (Figure 13.8). The TLC plate is immersed into the first developing solvent (chloroform/methanol/28% aqueous ammonia; 65:25:5, v/v/v) and is allowed to reach 1 cm from the top of the plate. After drying the TLC plate by flushing N_2 over the plate in a TLC box, the plate is developed at right angle to the first direction; see Figure 13.8. The solvent system used is chloroform/acetone/methanol/acetic acid/

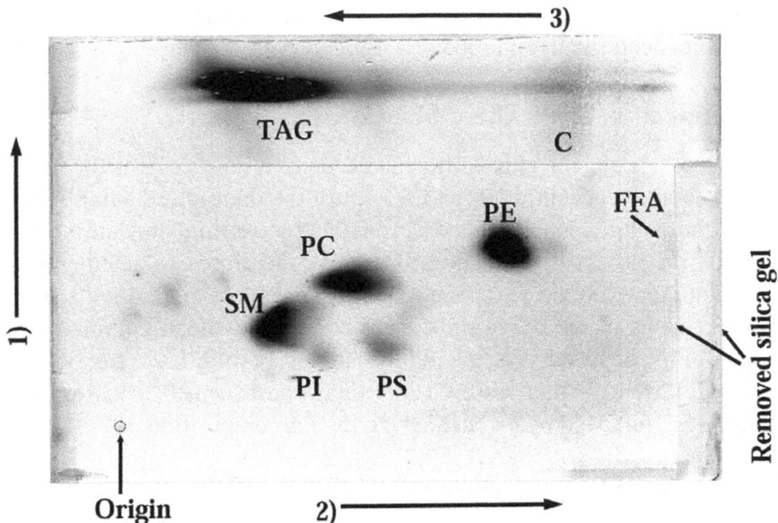

FIGURE 13.8 A typical separation of total milk fat obtained using a three-directional TLC system. The three solvent systems were chloroform/methanol/28% aqueous ammonia (65:25:5) in the first direction (1); chloroform/acetone/methanol/acetic acid/water (50:20:10:15:5) in the second direction (2); and hexane/diethyl ether/acetic acid (85:15:1) in the third direction (3). The separated lipid classes are triacylglycerol (TAG), cholesterol (C), phosphatidylcholine (PC), phosphatidylinositol (PI), phosphatidylethanolamine (PE), sphingomyelin (SM), phosphatidylserine (PS) and free fatty acids (FFAs).

water (50:20:10:15:5, v/v/v/v/v). The plate is developed until the solvent front reaches about 2 cm from the top of the plate. The plate is dried again by flushing N_2 over the TLC plate. The bands are visualized after spraying with 2′7′-dichlorofluorescein and viewed under UV light. A line is drawn in the same direction as the second development, separating the resolved PLs and neutral lipids concentrated along the solvent front. A 2-cm-wide band of silica is removed above the solvent front of the PL region after the second development. By removing the silica, the third developing solvent is restricted to the neutral lipid region, which avoids interference of the developed PLs. The TLC plate is then developed in the reverse direction from the second development using hexane/diethyl ether/acetic acid (85:15:1, v/v/v) as the solvent system. The TLC plate is dried again and visualized under UV light after spraying with 2′7′-dichlorofluorescein. Lipid classes are identified and scraped off for further analysis, or the plate is sprayed with H_2SO_4/methanol (1:1) and charred.

Figure 13.8 shows a typical separation obtained from total milk fat by three-directional TLC that resolves all the lipid classes of total milk fat. The relative elution order of the lipid classes was established previously.[101] The three-directional TLC separation allows for easy identification of all the different neutral and PL classes using only a small amount of milk fat per plate. Identification of the minor lipids requires accumulation of material from two or three plates. For consistency and reproducibility, the TLC plates should be dry and maintained at low humidity. The minor PL (PS and PI) may be difficult to visualize by UV light, in which case larger amounts of total lipid should be applied. If only selected PLs are being investigated one can use a one-directional TLC developing solvent such as chloroform/methanol/acetic acid/water (65:43:1:3).

13.9.2 Separation of PL Using SPE Columns

One of the major problems of isolating the PL present in milk fat is the interference caused by the high concentration of TAG. Both the three-directional TLC separation[101] and the column separations described[62,104] were time consuming to extract and quantify the PL constituents in milk fat. We therefore evaluated and modified an SPE technique to obtain sufficient material to conduct qualitative and quantitative measurements of the PL in milk fat. The Supelco-Clean columns gave good separations of the neutral lipids, glycolipids and PL of milk fat. After column separation, the PLs were further resolved into their constituent PL classes using one-directional TLC followed by GC to analyze the FA composition.

13.9.2.1 Suggested SPE Columns Method

Total milk fat (0.5 mg) dissolved in chloroform is added onto a 3 mL Supelco-Clean LC-SI column (Supelco, Bellefonte, PA). Three different fractions are collected into separate round bottom flasks, using three different solvent systems: (1) 30 mL of chloroform for the neutral lipids; (2) 25 mL of acetone for glycolipids; and (3) 10 mL of chloroform/methanol (1:1, v/v) and 30 mL of methanol for PLs. It is necessary to check the glycolipid fraction for the presence of any PL and neutral lipids because they easily coelute with acetone.[98] Solvents from each fraction are removed using a rotary evaporator as described before. Each column fraction was checked by TLC using silica gel G plates and hexane/diethyl ether/acetic acid (85:15:1) as a developing solvent.

The different PL classes were separated by TLC using silica gel H plates and chloroform/methanol/acetic acid/water (65:43:1:3) as developing solvent. The three most abundant PLs in milk fat (PE, PC, SM) were well resolved, but PI and PS could not always be identified because their concentration was too low. The identified PLs were scraped off into 5 mL vials to which known amounts of internal standard (17:0 FAME) were added for quantitative purposes. All the lipids are methylated using $NaOCH_3$/methanol, except for SM, which was acid methylated using HCl/methanol to ensure complete derivatization of the N-acyl lipids. GC is performed for each isolated band using the same conditions as for total milk fat.

13.10 CONCLUSION

Changes in milk FA composition can improve the nutritional status of dairy fat with respect to several FAs recognized to have health benefits, e.g., omega-3 FA, vaccenic acid (11*t*-18:1) and 9c11*t*-CLA. Under- and overestimation or misidentification of the FA in milk fat often occurs if inappropriate methods are used. Therefore, there is a need to use proven methods to obtain reliable results of ruminant fats, particularly when the diets of ruminants were modified to increase or decrease certain FAs. Some diets will cause shifts in the rumen bacterial flora and produce fats with undesirable lipid components. Complementary methods are presented in this chapter to provide precise and comprehensive analyses of dairy products. These proposed methodologies can also be applied for the lipid analysis of different matrices such as meat.[17] A reliable solvent system is described to extract total lipids using a chloroform/methanol/water mixture. Both base and acid catalyzed methylations should be conducted, the former to prevent CLA isomerization and the latter to provide the total lipid composition. The use of 100 m high-capillary GC columns is mandatory for the complete analysis of milk fat. Analysis of the total FAs in milk fat may require more than one GC determination, either at different sample loads or temperature programs. To identify and separate the overlapping monounsaturated FAs, a combination of Ag^+-TLC and isothermal GC is necessary. It is now well known that a complete analysis of the CLA isomers requires both GC and Ag^+-HPLC separations. The different CLA isomers have different biological or physiological properties and therefore should be determined. The total PL from milk fat can be effectively concentrated using SPE columns, separated by TLC and quantitated and analyzed by GC. Analysis of the PL content and composition in milk fat remains a challenge because of their relative low content, but this analysis is necessary to provide a complete composition of milk fat.

ACKNOWLEDGMENTS

The authors wish to thank the Dairy Farmers of Ontario and Consejo Nacional de Ciencia y Tecnologia from Mexico for C. Cruz-Hernandez. Contribution number #290 from the Food Research Program, AAFC.

REFERENCES

1. Walstra, P., Geurts, T.J., Noomen, A., Jellema, A. and van Boekel, M.A.J.S., *Dairy Technology: Principles of Milk Properties and Processes*, Marcel Dekker Inc., New York, 1999.
2. Varnam, A.H. and Sutherland, J.P., *Milk and Milk Products, Technology, Chemistry and Microbiology*, Volume l, Food Products Series, Chapman and Hall, London, 1994.
3. Christie, W.W., Composition and structure of milk lipids, in *Advanced Dairy Chemistry*, Volume 2, Fox, P.F., Ed., 2nd ed., Chapman and Hall, London, 1995, pp. 1–36.
4. Beaulieu, A.D. and Palmquist, D.L., Differential effects of high fat diets on fatty acid composition in milk of Jersey and Holstein cows, *J. Dairy Sci.*, 78, 1336–1344, 1995.
5. Chilliard, Y., Ferlay, A., Mansbridge, R.M. and Doreau, M., Ruminant milk fat plasticity: Nutritional control of saturated, polyunsaturated, *trans* and conjugated fatty acids, *Ann. Zootech.*, 49, 181–205, 2000.
6. Hillbrick, C. and Augustin, M.A., Milk fat characteristics and functionality opportunities for improvement, *Aust. J. Dairy Technol.*, 57, 45–51, 2002.
7. Gurr, M.I., Harwood, J.L. and Frayn, K.N., *Lipid Biochemistry, An Introduction*, 5th ed., Blackwell Science, Malden, MA, 2002.
8. Mensink, R.P., Zock, P.L., Kester, A.D.M. and Katan, M.B., Effects of dietary fatty acids and carbohydrates on the ratio of serum total to HDL cholesterol and on serum lipids and apolipoproteins: A meta-analysis of 60 controlled trials, *Am. J. Clin. Nutr.*, 77, 1146–1155, 2003.

9. Miller, G.D., Jarvis, J.K. and McBean, L.D., The importance of milk and milk products in the diet, in *Handbook of Dairy Foods and Nutrition*, 2nd ed., CRC Press, Boca Raton, FL, 2000, pp. 155–191.
10. Pariza, M.W., Park, Y. and Cook, M.E., The biologically active isomers of conjugated linoleic acid, *Prog. Lipid Res.*, 40, 283–298, 2001.
11. Parodi, P.W., Milk fat in human nutrition, *Austr. J. Dairy Technol.*, 59, 3–59, 2004.
12. Belury, M.A., Dietary conjugated linoleic acid in health: Physiological effects and mechanisms of action, *Ann. Rev. Nutr.*, 22, 505–531, 2002.
13. Parodi, P.W., Conjugated linoleic acid in foods, in *Advances in Conjugated Linoleic Acid Research,* Volume 2., Sébédio, J.-L., Christie, W.W. and Adlof, R., Eds., AOCS Press, Champaign, IL, 2003, pp. 101–122.
14. Lock, A.L. and Bauman, D.E., Modifying milk fat composition of dairy cows to enhance fatty acids beneficial to human health. *Lipids*, 39, 1197–1206, 2004.
15. Precht, D. and Molkentin, J., Identification and quantitation of cis/*trans* C16:1 and C17:1 fatty acid positional isomers in German human milk lipids by thin-layer chromatography and gas chromatography/mass spectrometry, *Eur. J. Lipid Sci. Technol.*, 102, 102–113, 2000.
16. Azizian, H. and Kramer, J.K.G., A rapid method for the quantification of fatty acids in fats and oils with emphasis on *trans* fatty acids using Fourier transform near infrared spectroscopy (FT–NIR), *Lipids*, 40, 855–867, 2005.
17. Cruz-Hernandez, C., Kramer, J.K.G., Kraft, J., Santercole, V., Or-Rashid, M., Deng, Z., Dugan, M.E.R., Delmonte, P. and Yurawecz, M.P., Systematic analysis of *trans* and conjugated linoleic acids in the milk and meat of ruminants, in *Advances in Conjugated Linoleic Acid Research*, Volume 3, Yurawecz, M.P., Kramer, J.K.G., Gudmundsen, O., Pariza, M.W. and Banni, S., Eds., AOCS Press, Champaign, IL, 2006, pp. 45–93.
18. Weggemans, R.M, Rudrum, M. and Trautwein, E.A., Intake of ruminant verses industrial *trans* fatty acids and risk of coronary heart disease—What is the evidence? *Eur. J. Lipid Sci. Technol.,* 106, 390–397, 2004.
19. Jensen, R.G. and Newburg, D.S., Bovine milk lipids, in *Handbook of Milk Composition*, Jensen, R.G. and Thompson M.P., Eds., Academic Press Inc., San Diego, 1995, pp. 543–575.
20. Kramer, J.K.G., Cruz-Hernandez, C., Deng, Z., Zhou, J., Jahreis, G. and Dugan, M.E.R., The analysis of conjugated linoleic acid and *trans* 18:1 isomers in synthetic and animal products, *Am. J. Clin. Nutr.*, 79 (Suppl.), 1137S–1145S, 2004.
21. Cruz-Hernandez, C., Deng, Z., Zhou, J., Hill, A.R., Yurawecz, M.P., Delmonte, P., Mossoba, M.M., Dugan, M.E.R. and Kramer, J.K.G., Methods for analysis of conjugated linoleic acids and trans-18:1 isomers in dairy fats by using a combination of gas chromatography, silver-ion thin-layer chromatography/gas chromatography and silver-ion liquid chromatography, *J. AOAC Internat.*, 87, 545–562, 2004.
22. Abu-Ghazaleh, A. A., Schingoethe, D. J. and Hippen, A. R., Conjugated linoleic acid and other beneficial fatty acids in milk fat from cows fed soybean meal, fish meal, or both, *J. Dairy Sci.*, 84, 1845–1850, 2001.
23. Mattos, R., Staples, C.R., Williams, J., Amorocho, A., McGuire, M.A. and Thatcher, W.W., Uterine, ovarian and production responses of lactating dairy cows to increasing dietary concentrations of menhaden fish meal, *J. Dairy Sci.*, 85, 755–764, 2002.
24. Petit, H.V., Dewhurst, R.J., Scollan, N.D., Proulx, J.G., Khalid, M., Haresign, W., Twagiramungu, H. and Mann, G.E., Milk production and composition, ovarian function and prostaglandin secretion of dairy cows fed omega-3 fats, *J. Dairy Sci.*, 85, 889–899, 2002.

25. Offer, N.W., Marsden, M., Dixon, J., Speake, B.K. and Thacker, F.E., Effect of dietary fat supplements on levels of n-3 polyunsaturated fatty acids, *trans* acids and conjugated linoleic acid in bovine milk, *Anim. Sci.*, 69, 613–625, 1999.
26. Baer, R.J., Ryali, J., Schingoethe, D.J., Kasperson, K.M., Donovan, D.C., Hippen, A.R. and Franklin, S.T. Composition and properties of milk and butter from cows fed fish oil, *J. Dairy Sci.*, 84, 345–353, 2001.
27. Chouinard, P.Y., Corneau, L., Butler, W.R., Chilliard, Y., Drackley, J.K. and Bauman, D.E., Effect of dietary lipid source on conjugated linoleic acid concentrations in milk fat, *J. Dairy Sci.*, 84, 680–690, 2001.
28. Donovan, D.C., Schingoethe, D.J., Baer, R.J., Ryali, J., Hippen, A.R. and Franklin, S.T., Influence of dietary fish oil on conjugated linoleic acid and other fatty acids in milk fat from lactating dairy cows, *J. Dairy Sci.*, 83, 2620–2628, 2000.
29. Shingfield, K.J., Ahvenjärvi, S., Toivonen, V., Ärölä, A., Nurmela, K.V.V., Huhtanen, P. and Griinari, J.M., Effect of dietary fish oil on biohydrogenation of fatty acids and milk fatty acid content in cows, *Anim. Sci.*, 77, 165–179, 2003.
30. Shingfield, K.J., Reynolds, C.K., Lupoli, B., Toivonen, V., Yurawecz, M.P., Delmonte, P., Griinari, J.M., Grandison, A.S. and Beever, D.E., Effect of forage type and proportion of concentrate in the diet on milk fatty acid composition in cows given sunflower oil and fish oil, *J. Anim. Sci.*, 80, 225–238, 2005.
31. Franklin, S.T., Martin, K.R., Baer, R.J.. Schingoethe, D.J. and Hippen, A.R., Dietary marine algae (Schizochytrium sp.) increases concentrations of conjugated linoleic, docosahexaenoic and *trans* vaccenic acids in milk of dairy cows, *J. Nutr.*, 129, 2048–2054, 1999.
32. Chouinard, P.Y., Corneau, L., Barbano, D.M., Metzger, L.E. and Bauman, D.E., Conjugated linoleic acids alter milk fatty acid composition and inhibit milk fat secretion in dairy cows, *J. Nutr.*, 129, 1579–1584, 1999.
33. Ip, C., Banni, S., Angioni, E., Carta, G., McGinley, J., Thompson, H.J., Barbano, D. and Bauman, D.A., Conjugated linoleic acid-enriched butter fat alters mammary gland morphogenesis and reduces cancer risk in rats, *J. Nutr.*, 129, 2135–2142, 1999.
34. French, P., Stanton, C., Lawless, F., O'Riordan, E.G., Monahan, F.J., Caffrey, P.J. and Moloney, A.P., Fatty acid composition, including conjugated linoleic acid, of intramuscular fat from steers offered grazed grass, grass silage, or concentrate-based diets, *J. Anim. Sci.*, 78, 2849–2855, 2000.
35. Baumgard, L.H., Corl, B.A., Dwyer, D.A., Saebo, A. and Bauman, D.E., Identification of the conjugated linoleic acid isomer that inhibits milk fat synthesis, *Am. J. Physiol. Regulatory Integrative Comp. Physiol.*, 278, R179–R184, 2000.
36. Bell, J.A. and Kennelly, J.J., Conjugated linoleic acid enriched milk: A designer milk with potential, *Adv. Dairy Technol.*, 13, 213–228, 2001.
37. Bell, J.A., Griinari, J.M. and Kennelly, J.J., Effect of safflower oil, flaxseed oil, monensin and vitamin E on concentration of conjugated linoleic acid in bovine milk fat, *J. Dairy Sci.* 89, 733–748, 2006.
38. Mackle, T.R., Kay, J.K., Auldist, M.J., McGibbon, A.K.H., Philpott, B.A., Baumgard, L.H. and Bauman, D.E., Effects of abomasal infusion of conjugated linoleic acid on milk fat concentration and yield from pasture-fed dairy cows, *J. Dairy Sci.*, 86, 644–652, 2003.
39. Molkentin, J. and Precht, D., Optimized analysis of *trans*-octadecenoic acids in edible fats, *Chromatographia*, 41, 267–272, 1995.

40. Kramer, J.K.G., Sehat, N., Dugan, M.E.R., Mossoba, M.M., Yurawecz, M.P., Roach, J.A.G., Eulitz, K., Aalhus, J.L., Schaefer, A.L. and Ku, Y., Distributions of conjugated linoleic acid (CLA) isomers in tissue lipid classes of pigs fed a commercial CLA mixture determined by gas chromatography and silver-ion high performance liquid chromatography, *Lipids*, 33, 549–558, 1998.
41. Kramer, J.K.G., Blackadar, C.B. and Zhou, J., Evaluation of two columns (60-m SUPELCOWAX 10 and 100-m CP Sil 88) for analysis of milk fat with emphasis on CLA, 18:1, 18:1, 18:2 and 18:3 isomers and short- and long-chain FA, *Lipids*, 38, 823–835, 2002.
42. Precht, D. and Molkentin, J., Overestimation of linoleic acid and *trans*-C18:2 isomers in milk fats with emphasis on *trans* Δ9, *trans* Δ12-octadecadienoic acid, *Milchwissenschaft*, 58, 30–34, 2003.
43. Precht, D., Molkentin, J., McGuire, M.A., McGuire, M.K. and Jensen, R.G., Overestimates of oleic and linoleic acid contents in materials containing *trans* fatty acids and analyzed with short packed gas chromatographic columns, *Lipids*, 36, 213–216, 2001.
44. Wolff, R.L. and Precht, D., A critique of 50-m CP Sil 88 capillary columns used alone to assess *trans*-unsaturated FA in foods: The case of the TRANSFAIR study, *Lipids*, 37, 627–629, 2002.
45. Mossoba, M.M., McDonald, R.E., Roach, J.A.G., Fingerhut, D.D., Yurawecz, M.P. and Sehat, N., Spectral confirmation of *trans* monounsaturated C_{18} fatty acid positional isomers, *J. Am. Oil Chem. Soc.*, 74, 125–130, 1997.
46. Wolff, R.L. and Precht, D., Comments on the resolution of individual trans-18:1 isomers by gas–liquid chromatography, *J. Am. Oil Chem. Soc.*, 75, 421–422, 1998.
47. Precht, D. and Molkentin, J., C18:1, C18:2 and C18:3 *trans* and *cis* fatty acid isomers including conjugated *cis* Δ9, *trans* Δ11 linoleic acid (CLA) as well as total fat composition of German human milk lipids, *Nahrung*, 43, 233–244, 1999.
48. Kramer, J.K.G., Cruz-Hernandez, C. and Zhou, J., Conjugated linoleic acids and octadecenoic acids: Analysis by GC, *Eur. J. Lipid Sci. Technol*., 103, 600–609, 2001.
49. Wolff, R.L., Analysis of alpha-linolenic acid geometrical isomers in deodorized oils by capillary gas-liquid chromatography on cyanoalkyl polysiloxane stationary phases: A note of caution, *J. Am. Oil Chem. Soc.*, 71, 907–909, 1994.
50. Yurawecz, M.P., Roach, J.A.G., Sehat, N., Mossoba, M.M., Kramer, J.K.G., Fritsche, J., Steinhart, H. and Ku, Y., A new conjugated linoleic acid isomer, 7*trans*, 9*cis*-octadecadienoic acid, in cow milk, cheese, beef and human milk and adipose tissue, *Lipids*, 33, 803–809, 1998.
51. Bauman, D.E. and Griinari, J.M., Nutritional regulation of milk fat synthesis, *Ann. Rev. Nutr*., 23, 203–227, 2003.
52. Turpeinen, A.M., Mutanen, M., Aro, A., Salminen, I., Basu, S., Palmquist, D.L. and Griinari, J.M., Bioconversion of vaccenic acid to conjugated linoleic acid in humans, *Am. J. Clin. Nutr*., 76, 504–510, 2002.
53. Kraft, J., Collomb, M., Möckel, P., Sieber, R. and Jahreis, G., Differences in CLA isomer distribution of cow's milk lipids, *Lipids,* 38, 657–664, 2003.
54. Kramer, J.K.G., Cruz-Hernandez, C., Or-Rashid, M. and Dugan, M.E.R., The use of total *trans*-11 containing FA, rather than total (A)n–7" FA, is recommended to assess the content of FA with a positive health image in ruminant fats, *Lipids,* 39, 693–695, 2004.
55. Griinari, J.M., Dwyer, D.A., McGuire, M.A., Bauman, D.E., Palmquist, D.L. and Nurmela, K.V.V., *Trans*-octadecenoic acids and milk fat depression in lactating dairy cows, *J. Dairy Sc.,* 81, 1251–1261, 1998.

56. Piperova, L.S., Teter, B.B., Bruckental, I., Sampugna, J., Mills, S.E., Yurawecz, M.P., Fritsche, J., Ku, K. and Erdman, R.A., Mammary lipogenic enzyme activity, *trans* fatty acids and conjugated linoleic acids are altered in lactating dairy cows fed a milk fat-depressing diet, *J. Nutr.*, 130, 2568–2574, 2000.
57. Fumiss, B.S., Hannaford, A.J., Rogers, V., Smith, P.W.G. and Tatchell, A.R., *Vogel's Textbook of Practical Organic Chemistry*, 4th ed., ELBS, London, 1984.
58. Folch, J., Lees, M. and Sloane Stanley, G.H., A simple method for the isolation and purification of total lipids from animal tissues, *J. Biol. Chem.*, 226, 497–509, 1957.
59. Bligh, E.G. and Dyer, W.J., A rapid method of total lipid extraction and purification, *Can. J. Biochem. Physiol.*, 37, 911–917, 1959.
60. Hara, A. and Radin, N.S., Lipid extraction of tissues with a low-toxicity solvent, *Anal. Biochem.*, 90, 420–426, 1978.
61. Marmer, W.N. and Maxwell, R.J., Dry column method for the quantitative extraction and simultaneous class separation of lipids from muscle tissue, *Lipids,* 16, 365–371, 1981.
62. Maxwell, R.J., Mondimore, D. and Tobias, J., Rapid method for the quantitative extraction and simultaneous class separation of milk lipids, *J. Dairy Sci.*, 69, 321–325, 1986.
63. Collins, S.E., Jackson, M.B., Lammi-Keefe, C.J. and Jensen, R.G., The simultaneous separation and quantitation of human milk lipids, *Lipids,* 24, 746–749, 1989.
64. Kramer, J.K.G. and Zhou, J., Conjugated linoleic acids and octadecenoic acids: Extraction and isolation of lipids, *Eur. J. Lipid Sci. Technol.,* 103, 594–600, 2001.
65. Yurawecz M.P., Kramer, J.K.G. and Ku, Y., Methylation procedures for conjugated linoleic acid, in *Advances in Conjugated Linoleic Acid Research*, Volume. I, Yurawecz, M.P., Mossoba, M.M., Kramer, J.K.G., Pariza, M.W. and Nelson, G.J., Eds., AOCS Press, Champaign, IL, 1999, pp. 64–82.
66. Winterfeld, M. and Debuch, H., Die Lipoide einiger Gewebe und Organe des Menschen, *Hoppe-Seyler Z. Physiol. Chem.*, 345, 11–21, 1966.
67. Kramer, J.K.G., Fellner, V., Dugan, M.E.R., Sauer, F.D., Mossoba, M.M. and Yurawecz, M.P., Evaluating acid and base catalysts in the methylation of milk and rumen fatty acids with special emphasis on conjugated dienes and total *trans* fatty acids, *Lipids,* 32, 1219–1228, 1997.
68. Yurawecz, M.P., Hood, J.K., Roach, J.A.G., Mossoba, M.M., Daniels, D.H, Ku, Y., Pariza, M.W. and Chin, S.F., Conversion of allylic hydroxy oleate to conjugated linoleic acid and methoxy oleate by acid-catalyzed methylation procedures, *J. Am. Oil Chem. Soc.,* 71, 1149–1155, 1994.
69. Wolff, R.L. and Bayard, C.C., Improvement in the resolution of individual *trans*-18:1 isomers by capillary gas-liquid chromatography: Use of a 100-m CP Sil 88 column, *J. Am. Oil Chem. Soc.,* 72, 1197–1201, 1995.
70. Wolff, R.L., Bayard, C.C. and Fabien, R.J., Evaluation of sequential methods for the determination of butterfat fatty acid composition with emphasis on trans-18:1 acids. Application to the study of seasonal variations in French butters, *J. Am. Oil Chem. Soc.*, 72, 1471–1483, 1995.
71. Chardigny, J.-M., Wolff, R.L., Mager, E., Bayard, C.C., Sébédio, J.-L., Martine, L. and Ratnayake, W.M.N., Fatty acid composition of French infant formulas with emphasis on the content and detailed profile of *trans* fatty acids, *J. Am. Oil Chem. Soc.,* 73, 1595–1601, 1996.
72. Ulberth, F., Gabernig, R.G. and Schrammel, F., Flame-ionization detector response to methyl, ethyl, propyl and butyl esters of fatty acids, *J. Am. Oil Chem. Soc.,* 76, 263–266, 1999.

73. Christie, W.W., A simple procedure for rapid transmethylation of glycerolipids and cholesterol esters, *J. Lipid Res.*, 23, 1072–1075, 1982.
74. Precht, D. and Molkentin, J., Rapid analysis of the isomers of trans-octadecenoic acid in milk fat, *Int. Dairy J.*, 6, 791–809, 1996.
75. Chilliard, Y., Ferlay, A. and Doreau, M., Effect of different types of forages, animal fats or marine oils in cow's diet on milk fat secretion and composition, especially conjugated linoleic acid (CLA) and polyunsaturated fatty acids, *Livestock Prod. Sci.*, 70, 31–48, 2001.
76. Dobson, G., Christie, W.W. and Nikolova-Damyanova, B., Silver ion chromatography of lipids and fatty acids, *J. Chromatogr.*, B, 671, 197–222, 1995.
77. Nikolova-Damyanova, B., Christie, W.W. and Herslöf, B., Mechanistic aspects of fatty acid retention in silver ion chromatography, *J. Chromatogr.*, A, 749, 47–54, 1996.
78. Morris, L.J., Separations of lipids by silver ion chromatography, *J. Lipid Res.*, 7, 717–732, 1966.
79. Nikolova-Damyanova B., Herslof, B.G. and Christie, W.W., Silver ion high-performance liquid chromatography of derivatives of isomeric fatty acids, *J. Chromatogr.*, 609, 133–140, 1992.
80. Delmonte, P., Kramer, J.K.G., Banni, S. and Yurawecz, M.P., New developments in silver ion and reverse phase HPLC of CLA, in *Advances in Conjugated Linoleic Acid Research*, Volume. 3, Yurawecz, M.P., Kramer, J.K.G., Gudmundsen, O., Pariza, M.W. and Banni, S., Eds., AOCS Press, Champaign, IL, 2006, pp. 95–118.
81. Abu-Ghazaleh, A.A., Schingoethe, D.J., Hippen, A.R. and Whitlock, L.A., Feeding fish meal and extruded soybeans enhances the conjugated linoleic acid (CLA) content of milk, *J. Dairy Sci.*, 85, 624–631, 2002.
82. Precht, D. and Molkentin, M., Recent trends in the fatty acid composition of German sunflower margarines, shortenings and cooking fats with emphasis on individual C16:1, C18:1, C18:2, C18:3 and C20:1 *trans* isomers, *Nahrung*, 44, 222–228, 2000.
83. Sehat, N., Kramer, J.K.G., Mossoba, M.M., Yurawecz, M.P., Roach, J.A.G., Eulitz, K., Morehouse, K.M. and Ku, Y., Identification of conjugated linoleic acid isomers in cheese by gas chromatography, silver ion high performance liquid chromatography and mass reconstructed ion profiles. Comparison of chromatographic elution sequences, *Lipids,* 33, 963–971, 1998.
84. Kramer, J.K.G., Sehat, N., Fritsche, J., Mossoba, M.M., Eulitz, K., Yurawecz, M.P. and Ku, Y., Separation of conjugated linoleic acid, in *Advances in Conjugated Linoleic Acid Research*, Volume. I, Yurawecz, M.P., Mossoba, M.M., Kramer, J.K.G., Pariza, M.W. and Nelson, G.J., Eds., AOCS Press, Champaign, IL, 1999, pp. 83–109.
85. Eulitz, K., Yurawecz, M.P., Sehat, N., Fritsche, J., Roach, J.A.G., Mossoba, M.M., Kramer, J.K.G., Adlof, R.O. and Ku, Y., Preparation, separation and confirmation of the eight geometrical cis/*trans* conjugated linoleic acid isomers 8,10- through 11,13-18:2, *Lipids,* 34, 873–877, 1999.
86. Adlof, R.O and Lamm, T., Fractionation of *cis*- and trans-oleic, linoleic and conjugated linoleic fatty acid methyl esters by silver ion high-performance liquid chromatography, *J. Chromatogr.*, A., 799, 329–332, 1998.
87. Sehat, N., Yurawecz, M.P., Roach, J.A.G., Mossoba, M.M., Kramer, J.K.G. and Ku, Y., Silver-ion high-performance liquid chromatographic separation and identification of conjugated linoleic acid isomers, *Lipids,* 33, 217–221, 1998
88. Delmonte, P., Roach, J.A.G., Mossoba, M.M., Losi, G. and Yurawecz, M.P., Synthesis, isolation and GC analysis of all the 6,8– to 13,15–cis/*trans* conjugated linoleic acid isomers, *Lipids,* 39, 185–191, 2004.

89. Delmonte, P., Kataoka, A., Corl, B.A., Bauman, D.E. and Yurawecz, M.P., Relative retention order of all isomers of cis/*trans* conjugated linoleic acid FAME from the 6,8– to 13,15–positions using silver ion HPLC with two elution systems, *Lipids,* 40, 509–514, 2005.
90. Roach, J.A.G., Yurawecz, M.P., Kramer, J.K.G., Mossoba, M.M., Eulitz, K. and Ku, Y., Gas chromatography high resolution selected-ion mass spectrometric identification of trace 21:0 and 20:2 fatty acids eluting with conjugated linoleic acid isomers, *Lipids,* 35, 797–802, 2000.
91. Sehat, N., Rickert, R., Mossoba, M.M., Kramer, J.K.G., Yurawecz, M.P., Roach, J.A.G., Adlof, R.O., Morehouse, K.M., Fritsche, J., Eulitz, K.D., Steinhart, H. and Ku, Y., Improved separation of conjugated linoleic acid methyl esters by silver-ion high-performance liquid chromatography, *Lipids,* 34, 407–413, 1999.
92. Erickson, M.C., Chemistry and function of phospholipids, in *Food Lipids, Chemistry, Nutrition and Biotechnology*, 2nd ed. Revised, Akoh, C.C. and Min, D.B., Eds., Marcel Dekker, New York, 2002, pp. 41–62.
93. Morrison, W.R., Jack, E.L. and Smith, L.M., Fatty acids of bovine milk glycolipids and phospholipids and their specific distribution in the diacylglycerophospholipids, *J. Am. Oil Chem. Soc.,* 42, 1142–1147, 1965.
94. Gentner, P.R., Bauer, M. and Dieterich, I., Thin-layer chromatography of phospholipids, Separation of major phospholipid classes of milk without previous isolation from total lipid extracts, *J. Chromatogr.*, A., 206, 200–204, 1981.
95. Kennerly, D.A., Improved analysis of species of phospholipids using argentation thin-layer chromatography, *J. Chromatogr.*, A. 363, 462–467, 1986.
96. Bitman, J., Wood, D.L., Mehta, N.R., Hamosh, P. and Hamosh, M., Comparison of the phospholipid composition of breast milk from mothers of term and preterm infants during lactation, *Am. J. Clin. Nutr.*, 40, 1103–1119, 1984.
97. Jensen, R.G., The composition of bovine milk lipids: January 1996 to December 2000, *J. Dairy Sci.*, 85, 295–350, 2002.
98. Christie, W.W., *Lipid Analysis. Isolation, Separation, Identification and Structural Analysis of Lipids*, 3rd ed., The Oily Press, Bridgwater, UK, 2003.
99. Hamilton, J.G. and Comai, K., Rapid separation of neutral lipids, free fatty acids and polar lipids using prepacked silica Sep-Pak columns, *Lipids,* 23, 1146–1149, 1988.
100. Christie, W.W., *High-Performance Liquid Chromatography and Lipids. A Practical Guide,* Pergamon Press, Oxford, UK, 1987.
101. Kramer, J.K.G., Fouchard, R.C. and Farnworth, E.R., A complete separation of lipids by three-directional thin layer chromatography, *Lipids,* 18, 896–899, 1983.
102. Ruiz-Gutiérrez, V. and Pérez-Camino, M.C., Update on solid-phase extraction for the analysis of lipid classes and related compounds, *J. Chromatogr.,* A., 885, 321–341, 2000.
103. Juaneda, P. and Rocquellin, G., Rapid and convenient separation of phospholipids and non phosphorous lipids from rat heart using silica cartridges, *Lipids,* 20, 40–41, 1985.
104. Pietsch, A. and Lorenz, R.L., Rapid separation of the major phospholipid classes on a single aminopropyl cartridge, *Lipids,* 28, 945–947, 1993.

14 The Use of Enzyme Test Kits for Teaching Lipid Chemistry*

Robert A. Moreau

CONTENTS

14.1 INTRODUCTION—A COMPARISON OF VARIOUS COMMERCIAL ENZYME TEST KITS

Many methods are available for the analysis of lipids and other natural products. These include various chromatographic techniques (gas, liquid and thin-layer chromatographies), various spectroscopic techniques (including NMR and infrared spectroscopy) and spectrometry. Although they are rarely used in research laboratories, enzyme-based assays are used extensively for routine analyses in clinical chemistry and in food analysis laboratories. Most of these methods are highly selective (due to the strict substrate specificities of most enzymes), accurate and convenient (the only instrument required for most enzymatic methods is a spectrophotometer or sometimes a fluorometer). Many of these enzymatic methods can be purchased as kits that contain one "cocktail" reagent that includes a premixed solution with all the necessary enzymes, buffers, cofactors and other compounds required to conduct the analyses. Some kits contain individual vials of enzymes, and other reagents and stock solutions of each must be prepared.

Enzyme test kits are available for the analysis of more than a hundred compounds.[1] Clinical chemistry laboratories (medical and veterinary) probably constitute the largest user of enzyme test kits, mostly for the analysis of various compounds

*Mention of trade names or commercial products in this publication is solely for the purpose of providing specific information and does not imply recommendation or endorsement by the U.S. Department of Agriculture.

in blood (serum and plasma). Food chemistry laboratories also employ a number of enzyme test kits, mostly for the analysis of carbohydrate-based materials.

When considering the analysis of lipids, enzyme test kits are also widely used in clinical laboratories but are seldom used in other research labs or in teaching labs. Enzyme test kits are available for four types of lipids: cholesterol, triglycerides, phospholipids, and free fatty acids (FFAs) (Table 14.1). Some of the more popular non-lipid enzyme test kits are also included in Table 14.1. American Organization of Analytical Chemists (AOAC) International has compiled an interesting table that compares the properties of more than 100 enzyme test kits.[1] Because many types of lipids include covalently bound glycerol and various sugars, there may be applications of some of these non-lipid enzyme test kits for the analysis of certain types of lipids (such as glycerol-lipids and glycolipids).

Enzyme test kits can be purchased from many sources, but when considering enzyme test kits for lipids, there are only five major manufacturers (Table 14.1). The shelf life of enzyme test kits can be as short as several months in refrigeration, mainly due to the short shelf life of some types of enzymes. Several professional organizations publish "official methods" for the analysis of certain specific types of compounds. The purpose for developing official methods is to ensure that various commercial laboratories all use the same specific method for the analysis of the same compound so that analytical results from various labs can be meaningfully compared. The major professional organization that develops official methods for the analysis of lipids is the American Oil Chemists Society (AOCS), which has more than 50 AOCS official methods, with new methods being introduced almost each year. It is interesting to note that the AOCS has no official methods that involve the use of enzyme test kits. Similarly, the other two major professional organizations (the AOAC International and the American Association of Cereal Chemists, AACC) that publish official methods for food analysis also do not have any that employ enzyme test kits for lipid analyses, but they do have official methods that employ enzyme test kits for dietary fiber, beta glucans, glucose, starch and other types of analyses (Table 14.1).

Although I am not aware of any previous reports on this topic, I believe that enzyme test kits could be effectively used for teaching lipid chemistry and biochemistry. In 2003, our laboratory published a paper[2] in which we evaluated a commercial enzyme-based serum cholesterol test kit for the analysis of phytosterols (plant sterols) and phytostanols (natural plant sterols that are completely saturated, meaning that they contain no carbon–carbon double bonds). We concluded that the kit could be used for the analysis of phytosterols and phytostanols, but instead of a standard 5 min incubation required for cholesterol analysis, the incubation time should be increased to 60 min. Based on our experience with the cholesterol test kit, I believe that enzyme test kits for lipids can be effectively utilized for teaching lipid chemistry and biochemistry.

14.2 CHOLESTEROL TEST KITS

Most cholesterol enzyme test kits contain three enzymes, cholesterol esterase, cholesterol oxidase and peroxidase.[3] Cholesterol esterase is necessary only if the

TABLE 14.1
Some Examples of Enzyme Test Kits for the Analysis of Lipids and Other Biological Compounds

Lipids and Other Biological Compounds	Source	Instrument	Intended Samples	Official Method	Ref.
		Lipids			
Cholesterol, Total (Infinity™)	Thermo Electron R–Biopharm	S	Blood Foodstuffs		3
Cholesterol, Total (Amplex(Red)	Invitrogen, Molecular Probes	F	Blood		4
Cholesterol, LDL, Direct	Thermo Electron Waco	S	Blood		5,6
Cholesterol, HDL (Infinity™)	Thermo Electron Waco	S	Blood		7,8
Triglycerides (Infinity™)	Thermo Electron	S	Blood		10
Phosphatidylcholine	Waco	S	Blood		11
Free fatty acids	Waco	S	Blood		
		Other Biological Compounds			
Acetic acid	Megazyme R–Biopharm	S	Foodstuffs		
Dietary fiber (Total)	Megazyme	S	Foodstuffs	AOAC, AACC	
Ethanol	R–Biopharm	S	Foodstuffs, beverages		
Beta glucan	Megazyme	S	Foodstuffs	AOAC, AACC	
Glucose	Sigma Megazyme R–Biopharm	S	Foodstuffs, blood	AOAC, AACC	
Glycerol	Megazyme R–Biopharm	S	Foodstsuffs		
Starch (total)	Megazyme R–Biopharm	S	Foodstuffs	AOAC	
Sucrose	R–Biopharm	S	Foodstuffs		

S = Spectrophotometer (colorimeter), F = fluorometer

sample contains sterol esters (usually fatty acid esters), in which case it hydrolyzes the ester bond so total cholesterol (the sum of free plus esterified cholesterol) can be measured. The second and key enzyme utilized by all cholesterol enzyme test kits is cholesterol oxidase, which oxidizes cholesterol to a ketone and concomitantly generates H_2O_2 (Figure 14.1). The amount of H_2O_2 produced in the cholesterol oxidase

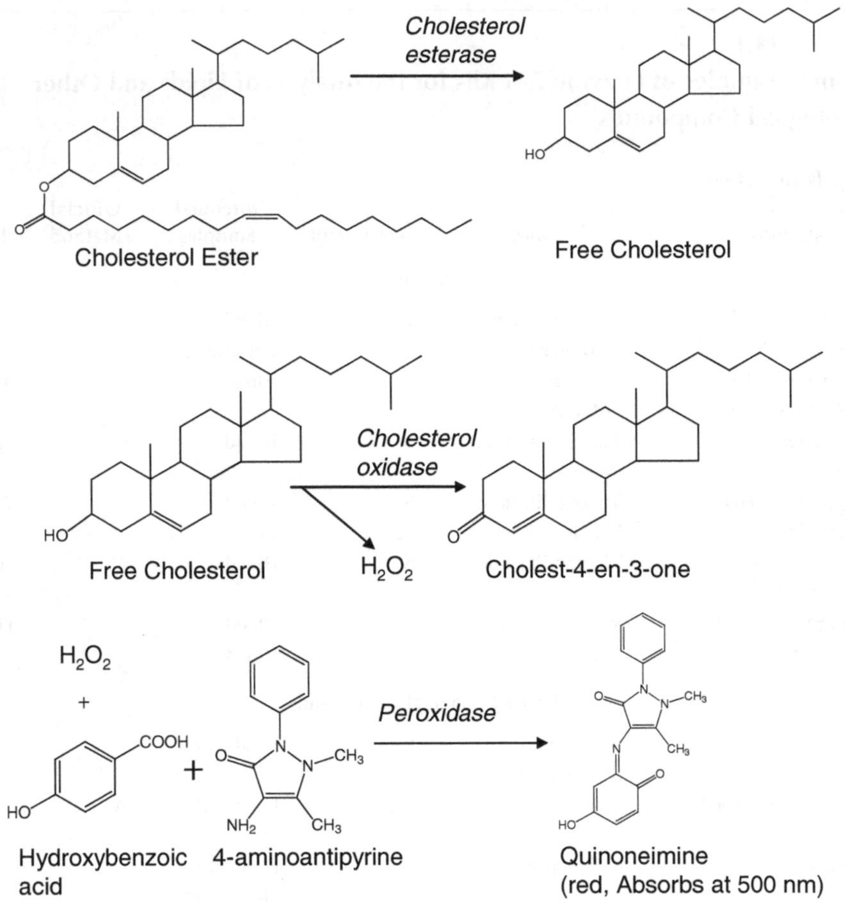

FIGURE 14.1 The enzymatic reactions that occur in an Infinity Cholesterol Test Kit.

reaction is quantified by utilizing it in a reaction that catalyzes the reaction H_2O_2 + 1 hydroxybenzoic acid + 4-aminoantipyrine to quinoneimine dye, which can be quantified in a spectrophotometer by monitoring absorbance at 500 nm (Figure 14.2).

In addition to the above spectrophotometric enzymatic method for analyzing total cholesterol, a second type of cholesterol enzymatic method, which utilizes a fluorometer, is also available and is more sensitive.[4] The fluorometric method employs the same three enzymes as the spectrophotometric method, but for the peroxidase reaction, H_2O_2 is quantified by reacting it with Amplex™ Red; the product is Resorufin, which fluoresces with an excitation maxima at 560 nm and an emission maxima at 590 nm (Figure 14.3). For those teaching labs that have a fluorometer capable of measurement at these excitation and emission wavelengths (or a fluorescence microtiter plate reader), the fluorometric method may be an interesting option.

In addition to measuring total cholesterol using the above methods, all clinical laboratories also now measure low-density lipoprotein cholesterol (LDL-C, sometimes called "bad" cholesterol because, when its levels are elevated, a patient is

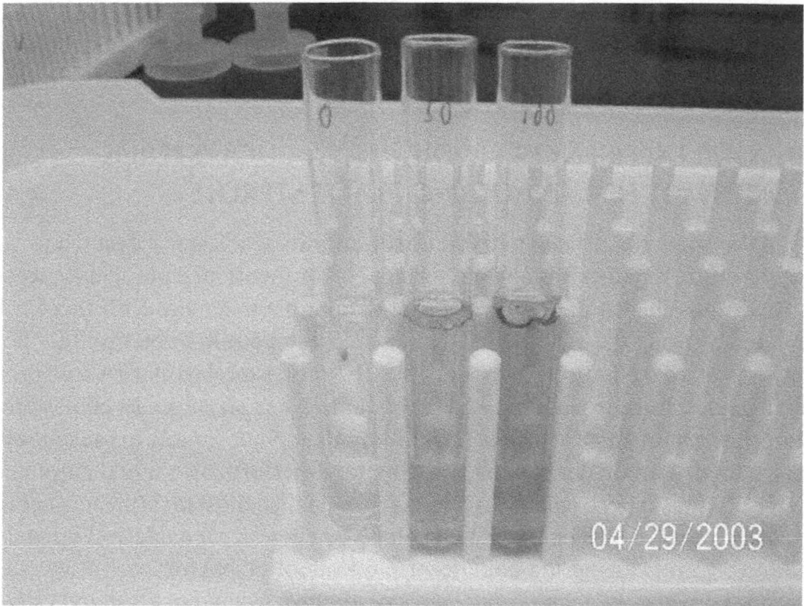

FIGURE 14.2 Photograph of three samples of cholesterol (0, 50 and 100 microliters of 0.005 M cholesterol, which corresponds to 0, 97 and 195 micrograms) after standard incubation in an Infinity Cholesterol Test Kit.

considered to be at a higher risk for developing cardiovascular disease) and high-density lipoprotein cholesterol (HDL-C, sometimes called "good" cholesterol because, when its levels are elevated, a patient is considered to be at a lower risk for developing cardiovascular disease). There are several enzymatic methods for

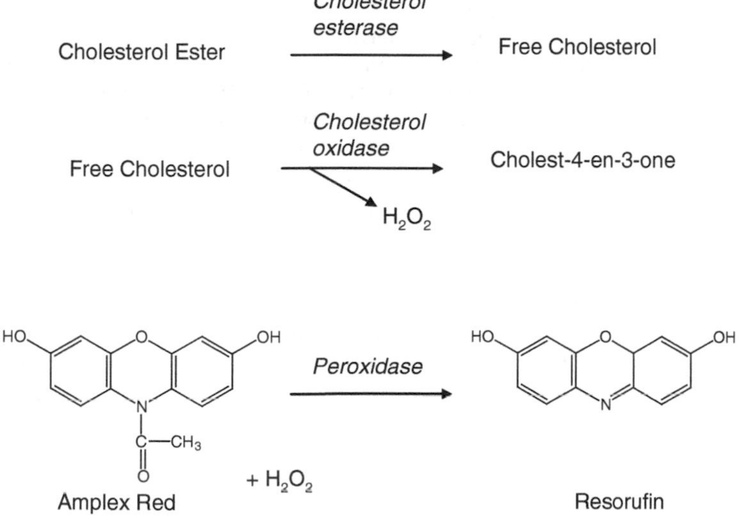

FIGURE 14.3 The enzymatic reactions that occur in an Amplex Red Cholesterol Test Kit.

measuring LDL-C and HDL-C, but interestingly, all of them employ the same enzymatic steps as for total cholesterol, but they also include other reagents that "block" either HDL-C or LDL-C.[5–8]

14.3 OUR EXPERIENCES MODIFYING A CHOLESTEROL TEST KIT FOR ANALYZING PLANT STEROLS

Several years ago we decided to try to use an Infinity Cholesterol Test Kit to quantitatively analyze phytosterols and phytostanols.[2] As mentioned previously, we found that it was possible to measure phytosterols and phytostanols with this kit but it required increasing the enzyme incubation step from 5 min to 60 min (Table 14.2). When we started the project we thought that the cholelsterol oxidase would probably oxidize phytosterols because they have the same 5-ene structure as cholesterol, but we were uncertain about whether it would oxidize phytostanols because they contain no carbon–carbon double bonds. Using high-performance liquid chromatography (HPLC) mass spectrometry, we demonstrated that sitostanol (the most common phytostanol) was indeed oxidized to a ketone by cholesterol oxidase (Figure 14.4). In addition to measuring the amount of phytosterol or phytostanol in the sample, we also used HPLC to quantify the levels of phytosteryl and phytostanyl esters, the levels of free phytosterol and free phytostanol, and the levels of "ketone 1" (produced from phytosterols) and "ketone 2" produced from phytostanols (Table 14.2). We demonstrated that the first enzyme in the Infinity Cholesterol Reagent, choles-

TABLE 14. 2
The Identification via HPLC–MS of the Substrates (Phytosteryl/Phytostanyl Esters or Free Phytosterols), Intermediates and Products Created during Incubation of Various Sterols and Stanols with Enzyme Mixtures

Substrate	Incubation time (min)	Steryl/stanyl fatty acyl ester	Steryl/stanyl ferulate ester	Free sterol	Free stanol	Ketone 1 (rt = 9 min)	Ketone 2 (rt = 3 min)	Abs 500
Cholesterol	0	0	0	100	0	0	0	0.000
	5							0.289
	10	0	0	0	0	98	2	0.307
Cholesterol: Oleate	0	100	0	0	0	0	0	0.000
	5							0.305
	10	0	0	0	0	99	1	0.311
Stigmasterol	0	0	0	100	0	0	0	0.000
	5	0	0	30	0	68	2	0.129
	15							0.193
	30							0.227
	60	0	0	0	0	99	1	0.258

(Continued)

TABLE 14.2
(Continued)

Substrate	Incubation time (min)	Steryl/ stanyl fatty acyl ester	Steryl/ stanyl ferulate ester	Free sterol	Free stanol	Ketone 1 (rt = 9 min)	Ketone 2 (rt = 3 min)	Abs 500
Stigmastanol	0	0	0	0	100	0	0	0.000
	5	0	0	0	17	2	80	0.176
	15							0.285
	30							0.316
	60	0	0	0	1	2	97	0.319
Take Control	0	98	0	2	0	0	0	0.000
	5	0	0	12	0	86	2	0.298
	15							0.339
	30							0.357
	60	0	0	0	0	97	3	0.357
Benecol	0	99	0	0	1	0	0	0.000
	5	0	0	0	16	0	84	0.315
	15							0.334
	30							0.340
	60	0	0	0	4	0	96	0.383
Cook Smart	0	98	0	2	0	0	0	0.000
	5	0	0	30	0	66	4	0.307
	15							0.389
	30							0.461
	60	0	0	1	0	97	2	0.541
Cholesterol	0	0	0	100	0	0	0	0.000
Success	5	0	0	27	0	58	15	0.230
	15							0.310
	30							0.337
	60	0	0	1	0	80	18	0.350
Corn fiber oil	0	55	41	4	0	0	0	0.000
	5	0	39	3	0	47	11	0.186
	15							0.215
	30							0.245
	60	0	36	1	0	50	13	0.275
Stigmastanyl–ferulate	0							0.000
Ferulate	60							0.009
Oryzanol	0							0.000
	60							0.006

Note: At each time point Abs 500 nm was also measured. Blank lines indicate that only Abs 500 nm was measured.

Source: Moreau, R.A., Powell, M.J. and Hicks, K.B., Evaluation of a commercial enzyme-based serum cholesterol test kit for the analysis of phytosterol and phytostanol products, *J. Ag. Food Chem.* 51, 6663–6667, 2003. With permission.

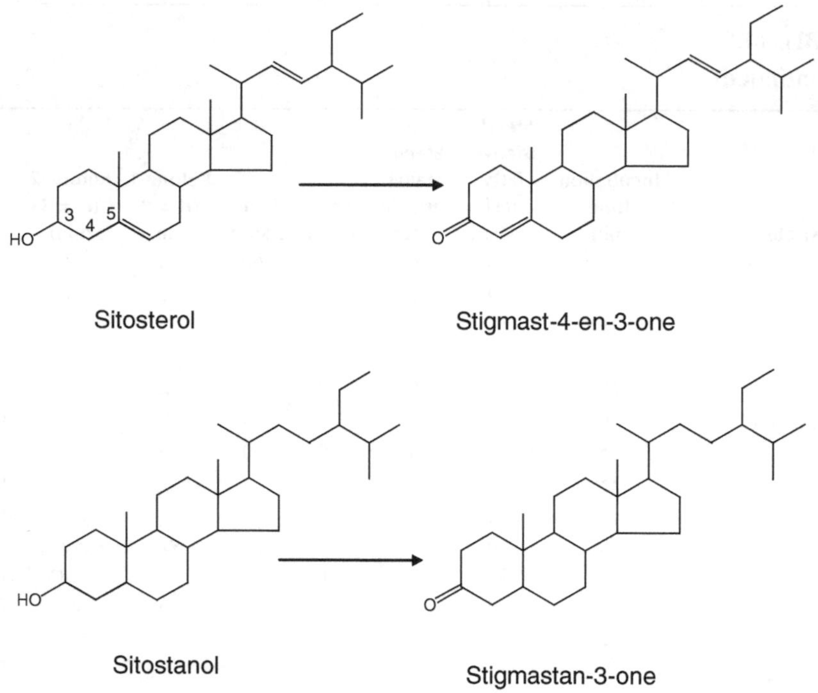

Sitosterol → Stigmast-4-en-3-one

Sitostanol → Stigmastan-3-one

FIGURE 14.4 The substrates and products of a common phytosterol (sitosterol) and phytostanol (sitostanol) when reacted in an Infinity Cholesterol Test Kit.

terol esterase, could hydrolyze the ester bonds on phytosteryl and phytostanyl esters (fatty acid esters) (Table 14.2). We also demonstrated that it could not hydrolyze the ester bond on phytosteryl or phytostanyl esters of ferulic acid, which are plentiful in rice bran oil (where they are called "oryzanol") and in corn fiber oil,[9] where they are mostly phytostanyl esters.

After we demonstrated that the Infinity Cholesterol Reagent could be used to analyze phytosterols and phytostanols, we then employed the modified method to measure their levels in three functional food products (Table 14.3). One of the products contained free unesterified sterols (Cholesterol Success), one contained phytostanyl esters (Benecol) and two contained phytosteryl esters (Take Control and Cook Smart). Cook Smart was test marketed in 2003 as a phytosterol-enriched cooking oil, but it is no longer commercially available.

14.4 TRIACYLGLYCEROL (TRIGLYCERIDE) TEST KITS

The Infinity Triglyceride Test Kit contains four enzymes: lipase, glycerol kinase, glycerolphosphate oxidase, and peroxidase.[10] The reactions that each of these enzymes catalyze are shown in Figure 14.5. As with most cholesterol test kits, the amount of H_2O_2 produced in the cholesterol oxidase reaction is quantified by utilizing it in a reaction that catalyzes the reaction H_2O_2 + 1 hydroxybenzoic acid + 4-aminoantipyrine to quinoneimine dye (which can be quantified in a spectrophotometer by monitoring absorbance at 500 nm). Because the lipase would also

TABLE 14.3

Estimation of Phytosterol Concentrations in Commercial Products Using the Infinity Cholesterol Reagent Method, Modified to Include a 60 Min Incubation

Product	Advertised mg Sterol (Ester or Free) Per Serving	Calculated Millimole Sterol Per Serving	Infinity Values Millimole Sterol Per Serving
Benecol	1500 mg[d]	2.20[a]	2.701 ± 0.012 (23% higher)
Take Control	1650 mg[e]	2.43[b]	2.751 ± 0.010 (13% higher)
Cook Smart	1120 mg[e]	1.65[b]	2.000 ± 0.022 (21% higher)
Cholesterol Success	900 mg[f]	2.17[c]	2.255 ± 0.002 (2% higher)

[a] Calculated assuming the fatty acyl phytostanyl esters are stigmastanol–oleate (MW 681.2); [b]calculated assuming the fatty acyl phytosteryl esters are sitosterol–oleate (MW 679.2); [c]calculated assuming that the free phytosterols are sitosterol (MW 414.7); [d]product has esterified phytostanols; [e]product has esterified phytosterols; [f]product has free (unesterified) phytosterols.

Source: Moreau, R.A., Powell, M.J. and Hicks, K.B., Evaluation of a commercial enzyme-based serum cholesterol test kit for the analysis of phytosterol and phytostanol products, *J. Ag. Food Chem.* 51, 6663–6667, 2003. With permission.

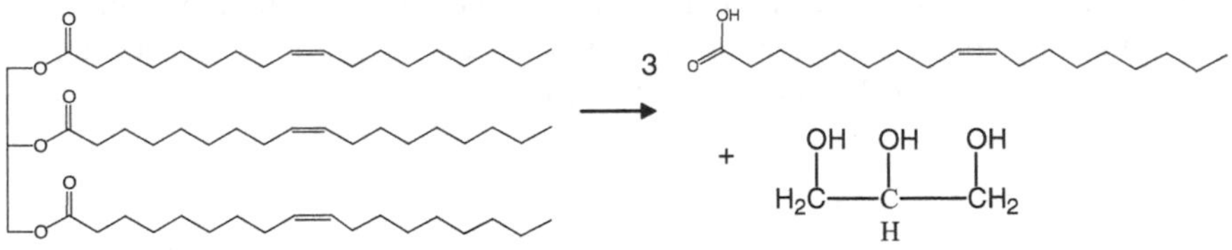

FIGURE 14.5 The enzymatic reactions that occur in an Infinity Triglyceride Test Kit.

produce glycerol from monoglycerides and diglycerides, this test kit would also measure the amounts of these compounds in a sample.

14.5 PHOSPHOLIPID (PHOSPHATIDLYCHOLINE) TEST KITS

The Waco Phosopholipid B Assay Kit contains three enzymes: phospholipase D, choline oxidase and peroxidase.[11] The reactions that each of these enzymes catalyze are shown in Figure 14.6. As with most cholesterol test kits and the triglyceride test kit, the amount of H_2O_2 produced in the cholesterol oxidase reaction is quantified by utilizing it in a reaction that catalyzes the reaction H_2O_2 + 1 hydroxybenzoic acid + 4-aminoantipyrine to quinoneimine dye (which can be quantified in a spectrophotometer by monitoring absorbance at 500 nm). Although this kit is marketed as a phospholipid test kit, it only detects one phospholipid, phosphatidylcholine. Also, because the phospholipase D would also generate choline from sphingomyelin and lyso-phosphatidylcholine, this test kit would also measure the amounts of these two other compounds in a sample.

14.6 FREE FATTY ACID TEST KITS

The Waco Non-Esterified Fatty Acid Assay Kit contains three enzymes: Acyl CoA synthetase, Acyl CoA oxidase and peroxidase. The reactions that each of these enzymes catalyze are shown in Figure 14.7. As with most of the previous test kits, the amount of H_2O_2 produced in the cholesterol oxidase reaction is quantified by utilizing it in a reaction that catalyzes the reaction H_2O_2 + 1 hydroxybenzoic acid + 4-aminoantipyrine to quinoneimine dye (which can be quantified in a spectrophotometer by monitoring absorbance at 500 nm).

Phosphatidylcholine (or sphingomyelin) (or lyso-phosphatidylcholine) —*Phospholipase D*→ Choline + Glycerolphosphate

Choline + O_2 —*Choline Oxidase*→ Betaine Aldehyde + H_2O_2

H_2O_2 + 4-aminoantipyrine + 3,5-dichloro-2-hydroxybenzene sulfate —*Peroxidase*→ Quinoneimine (red) 2 H_2O

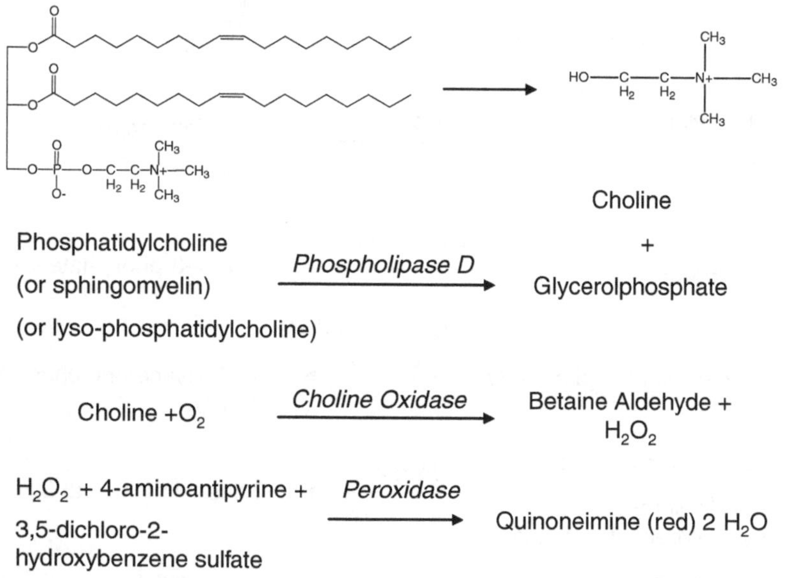

FIGURE 14.6 The enzymatic reactions that occur in a Waco Phospholipid B Test Kit.

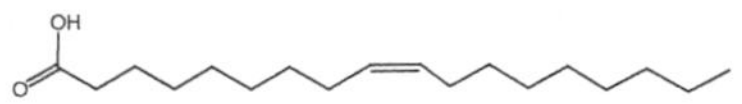

Fatty Acid + Coenzyme A $\xrightarrow{\textit{Acyl CoA Synthetase}}$ Fatty Acyl Co A

Fatty Acyl Co A + O_2 $\xrightarrow{\textit{Acyl Co A Oxidase}}$ trans-2,3-Dehydroacyl-CoA + H_2O_2

H_2O_2 + 4-aminoantipyrine + 3,5-dichloro-2-hydroxybenzene sulfate $\xrightarrow{\textit{Peroxidase}}$ Quinoneimine (red) 2 H_2O

FIGURE 14.7 The enzymatic reactions that occur in a Waco Free Fatty Acid Test Kit.

14.7 CONCLUSIONS—SOME THOUGHTS ON USING OTHER LIPID TEST KITS FOR TEACHING LIPID CHEMISTRY

Although they have received little attention except for clinical applications, I believe the above enzyme test kits could easily be utilized to teach basic and advanced principles of lipid chemistry and biochemistry. The only instrument needed in the teaching laboratory would be a simple spectrophotometer, colorimeter, or microtiter plate reader.

In a basic class, the Infinity Cholesterol Test Kit or the Infinity Triglyceride Test Kit may be useful because both of these are marketed as a "reagent cocktail" that contains all of the ingredients (enzymes, buffers, detergents, cofactors, etc.) necessary to conduct the assay. The student would need only to add an appropriate amount of sample to a test tube or microtiter plate, incubate for the appropriate time and temperature, and measure the absorbance in a spectrophotometer or microtiter plate reader.

In a more advanced class, the Amplex Red Cholesterol Test Kit may be more appropriate. This kit comes with separate vials that contain each of the individual enzymes and other components, so the students would need to prepare several stock reagents and pipet them using the prescribed procedures. One advantage to this type of kit is that it would be possible to measure both free cholesterol and cholesterol esters by choosing to add or omit the cholesterol esterase.

A suggestion for a possible laboratory experiment would be to measure the levels of phytosterols and tryglycerides in samples of conventional margarines and nutraceutical margarines such as Benecol and Take Control (Table 14.3). Because both margarines are readily available in most parts of the U.S. and in the U.K. (where the latter is marketed as Pro-Activ), analyzing these samples may make a good laboratory experiment for students. In addition, the students could also measure the levels of tryglycerides in these margarines using the triglyceride test kit described in the next section. Because all margarines contain some water—"light" margarines can contain as much as 70% water—measuring the levels of triglycerides in several types of margarines may make an interesting laboratory experiment. A suggested laboratory experiment is outlined in Scheme 14.1.

SCHEME 14.1
Suggested Experiment (Based on Procedure in Reference 2)

1. Purchase three soft (tub) margarines—one high-fat (~70%), one low-fat (~30%) and one phytosteryl ester (~ 10%) or phytostanyl ester enriched.
2. Remove the margarines from their containers and place them in new containers labeled A, B and C.
3. Purchase an Infinity Cholesterol Test Kit and an Infinity Triglyceride Test Kit.
4. Instruct the students to weigh 25 mg of each sample into a clean test tube and dissolve each in 5 ml of ethanol.
5. Instruct the students to pipet 10 microliters of each stock solution into two sets of test tubes.
6. With one set of tubes the students should pipet 1 ml of Infinity Cholesterol Reagent to each tube and with the other set they should pipet 1 ml of Infinity Triglyceride Reagent.
7. The tubes should be incubated shaking for 1 h at 37°C. If an incubator is not available, the experiment can still be conducted but the red color will develop at a slower rate.
8. After an hour, either the red color of the tubes can be observed visually (see Figure 14.2) or the absorbance 500 nm of each tube can be measured with a spectrophotometer or a colorimeter. For the Infinity Cholesterol Reagent and phytosterol or phytostanol margarines, the absorbance values should be similar to those in Table 14.2. With the Infinity Triglyceride Reagent, the instructor should conduct the experiment with several concentrations of triglycerides to be able to correlate the intensity of red color with amount of trigylyceride.
9. Based on the qualitative or quantitative results from step 8, the students should be able to determine the identity of the samples as high-fat (red with TAG Reagent), low-fat (pink with TAG Reagent) or phytosterol/phytostanyl spread (pink with Cholesterol Reagent).

REFERENCES

1. Anonymous, Biochemical Test Kits, www.aoac.org/testkits/kits-biochemistry.
2. Moreau, R.A., Powell, M.J. and Hicks, K.B., Evaluation of a commercial enzyme-based serum cholesterol test kit for the analysis of phytosterol and phytostanol products, *J. Ag. Food Chem.* 51, 6663–6667, 2003.
3. Anonymous, Infinity Cholesterol Reagent, http://www.thermo.com/eThermo/CMA/PDFs/Various/File_28607.pdf.
4. Anonymous, Amplex Red Cholesterol Assay Kit (A12216), http://probes.invitrogen.com/media/pis/mp12216.pdf.
5. Anonymous, Direct LDL Cholesterol Plus Reagent, http://www.thermo.com/eThermo/CMA/PDFs/Various/File_28701.pdf.

6. Maitra, A., Hirany, S.V. and Ishwaral, J., Comparison of two assays for measuring LDL cholesterol, *Clin. Chem.* 43, 1040–1047, 1997.
7. Anonymous, Infinity™ HDL Cholesterol Automated Reagent, http://www.thermo.com/eThermo/CMA/PDFs/Various/File_28494.pdf.
8. Lin, M.-J., Hoke, C. and B. Ettinger, Evaluation of homogeneous high-density lipoprotein cholesterol assay on a BM/Hitachi 747–200 Analyzer, *Clin. Chem.* 44, 1050–1052, 1998.
9. Moreau, R.A., Whitaker, B.D. and Hicks, K.B., Phytosterols, phytostanols and their conjugates in foods: Structural diversity, quantitative analysis and health-promoting uses, *Prog. Lipid Res.* 41, 457–500, 2002.
10. Anonymous, Infinity Triglycerides Reagent, http://www.thermo.com/eThermo/CMA/PDFs/Various/File_28372.pdf.
11. Grohganz, H., Ziroli, V., Massing, U. and Brandl, M., Quantification of various phosphatidylcholines in liposomes by enzymatic assay, *AAPS Pharm. Sci. Tech.* 63, 1–6, 2003.

15 Plastids as a Model System in Teaching Plant Lipid Metabolism*

*Salvatore A. Sparace***
and Kathryn F. Kleppinger-Sparace

CONTENTS

15.1 INTRODUCTION

Plastids are an extremely important and diverse group of organelles found in higher plant cells. As such, they occur in a variety of forms with a variety of specialized functions. Perhaps best known among these are chloroplasts, which occur in a variety of photosynthetic tissues and are engaged in the photosynthetic activities of plants (Halliwell, 1981; Malkin and Niyogi, 2000). The historical emphasis on chloroplasts often overshadows the important contributions that a number of nonphotosynthetic

*This work is dedicated to the memory of our teacher, our mentor and our friend John Brian Mudd (1929–1998). He was the first to ever teach us about chloroplasts and their role in plant lipid metabolism. But even more, by his example he showed us how to enjoy science and what it really meant to be good scientists and good teachers.

**Corresponding author

plastids make to the overall physiology of the plant that rival those of their photosynthetic counterparts. These include the chromoplasts of flowers and fruit that commonly provide the characteristic yellow, orange and red colors of these organs, the amyloplasts of tubers and other storage organs that are important in the synthesis and storage of starch and the leucoplasts of developing oilseeds involved in the synthesis of fatty acids for oil accumulation (Somerville et al., 2000; Neuhaus and Emes, 2000). All plastids are developmentally related to proplastids (Staehelin and Newcomb, 2000; Waters and Pyke, 2005). Proplastids are characteristically small diameter (0.2–1.0 μ) spherical colorless plastids with a poorly developed internal membrane system. They are commonly found in the young meristematic tissues of plants and give rise to leucoplasts, amyloplasts, chromoplasts and eventually chloroplasts (Staehelin and Newcomb, 2000; Waters and Pyke, 2005).

Despite the wide range of specializations in plastid form and function, all plastids have in common a number of primary metabolic or biosynthetic processes that are vital to the plant cell and the entire plant. Indeed, as a result of numerous studies of plastid function, plastids are emerging as the "biosynthetic powerhouse" of the plant cell (Neuhaus and Emes, 2000). In this regard, besides the processes already mentioned, plastids are almost universally involved in nitrogen and sulfur assimilation, which includes the reduction of nitrite to ammonia and its subsequent incorporation into amino acids via the glutamine synthase/glutamate synthase ("GS/GOGAT") cycle (Crawford et al., 2000; Tetlow et al., 2005); the activation and reduction of sulfate to sulfide and its incorporation into cysteine (Crawford et al., 2000); and finally the biosynthesis of isoprenoids and aromatic amino acids (Tetlow et al., 2005). All of these biosynthetic processes require a supply of metabolic energy and reduced carbon. Unlike the highly specialized chloroplasts that are capable of providing their own energy (ATP, NADPH) and reduced carbon intermediates, nonphotosynthetic plastids must rely, either directly or indirectly, on the cytosolic compartment for their carbon and energy requirements (Neuhaus and Emes, 2000). To facilitate this crucial interaction with the cytosolic compartment, plastids may have as many as 45 putative transporters variously situated in the membranes of the plastid envelope (Weber et al., 2005). Among these are the triose-phosphate/phosphate translocator; the glucose 6-phosphate/phosphate translocator; the phosphoenolpyruvate/phosphate translocator; the ATP/ADP translocator; the pentose phosphate, glucose and ADP-glucose transporters; and the dicarboxylate transporter family (Weber et al., 2005). Equally important, it is now known that plastids have essentially complete sets of enzymes of both the glycolytic and oxidative pentose phosphate pathways (Neuhaus and Emes, 2000; Tetlow et al., 2005). These pathways can provide a variety of key intermediates including the energy (ATP) and reducing power (NADH and NADPH) required for nitrogen and sulfur assimilation and fatty acid biosynthesis as well as the glucose-6-phosphate for starch biosynthesis, and the erythrose-4-phosphate and phosphoenolpyruvate for aromatic amino acid biosynthesis. Thus, it is clear that plastids can either obtain or synthesize essentially all of the raw materials they require to support their various activities, especially the acetyl-CoA needed for *de novo* fatty acid biosynthesis (Tetlow et al., 2005) and glycerolipid assembly (Joyard et al., 1998).

15.2 LIPID BIOSYNTHESIS IN PLANT CELLS

Lipids have a wide variety of functions in higher plant cells. These range from waterproofing and protection of the outer surfaces of plant tissues to participation in several signal transduction cascades. Among the various types of lipids found in plants, the glycerolipids make up the great majority and they have two main functions. First, they are the main class of structural lipids found in all membranes of the cell, and second, they serve as important storage reserves, especially in oil-storing tissues and seeds (Somerville et al., 2000). The biosynthesis of a full complement of complex glycerolipids for any given plant cell requires the metabolic interaction of three cellular organelles. Paramount among these are the plastids. As already mentioned, plastids are considered to be the only site of *de novo* fatty acid biosynthesis in the cell. As such, they are charged with the metabolic responsibility of providing a supply of fatty acids for glycerolipid assembly in the various cell compartments involved in membrane or storage lipid biosynthesis. Besides the plastid itself, other sites of glycerolipid assembly in the cell are the endoplasmic reticulum and its associated developing oil bodies, and the mitochondria (Joyard et al., 1998; Somerville et al., 2000).

Acetyl-CoA is the immediate precursor for *de novo* fatty acid biosynthesis. However, acetyl-CoA itself is not transported across the plastid envelope. Instead, a number of other metabolites that are readily transported into the plastid can be effectively metabolized to pyruvate, which in turn is converted to acetyl-CoA via a plastidic pyruvate dehydrogenase complex. These include the uptake of hexose and triose phosphates that are metabolized to pyruvate via the plastidic glycolytic pathway and malate that is converted to pyruvate via NADP-malic enzyme. Pyruvate can also be directly taken up from the cytosol via an as-yet-uncharacterized pyruvate transporter. Alternatively, photosynthetically formed phosphoglycerate in chloroplasts and its subsequent conversion to pyruvate via glycolytic reactions in the plastid is also feasible (Bao et al., 2000).

Whatever the source of pyruvate, it is likely that plastids can utilize multiple precursors *in vivo*, and that the physiological preference for a given precursor will depend on the nature or type of tissue from which plastids are isolated as well as the developmental stage of that tissue (Tetlow et al., 2005). However, acetate historically has been a standard precursor for the study of fatty acid biosynthesis and glycerolipid assembly in plastids. This owes largely to its supporting high rates of *in vivo* and *in vitro* lipid biosynthesis in chloroplasts (Roughan et al., 1976), leaves (Slack and Roughan, 1975) and fat-storing tissues (Stumpf and Barber, 1957). In this regard, mitochondrial-derived acetate is thought to be passively absorbed by plastids where it converts to acetyl-CoA by acetyl-CoA synthetase located in the stroma of the plastid (Kuhn et al., 1981) with ATP and CoA required as cosubstrates. Despite the extensive use of acetate as a precursor for the study of lipid metabolism, it is likely that pyruvate may be a physiologically more relevant precursor *in vivo* (Tetlow et al., 2005). Nevertheless, acetate remains an economical and effective substrate for the study of plastidic lipid metabolism.

Once acetyl-CoA is generated within the plastid, it is committed to lipid biosynthesis within the plastid by a multimeric acetyl-CoA carboxylase (ACCase), which

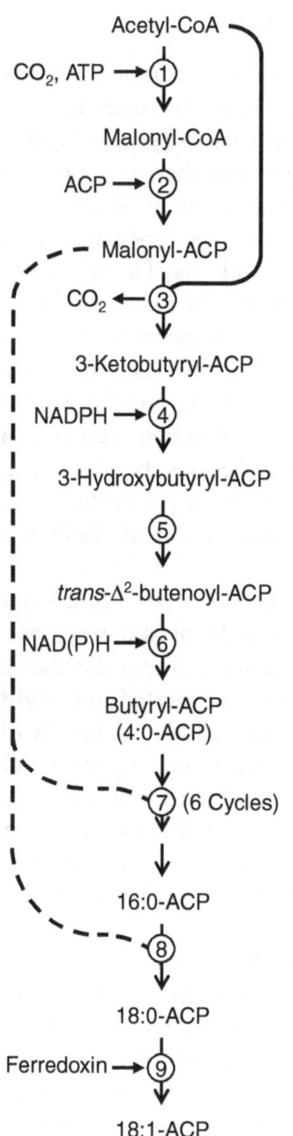

FIGURE 15.1 Enzymes and reactions of *de novo* fatty acid biosynthesis in plastids. 1 = acetyl-CoA carboxylase; 2 = malonyl-CoA : ACP transacylase; 3 = ketoacyl-ACP synthase III; 4 = 3-ketoacyl-ACP reductase; 5 = 3-hydroxyacyl-ACP dehydratase; 6 = 2,3-trans-enoyl-ACP reductase; 7 = ketoacyl-ACP synthase I; 8 = ketoacyl-ACP synthase II; 9 = stearoyl-ACP Δ9-desaturase; ACP = acyl carrier protein. (Adapted from Somerville et al., 2000.)

catalyzes its conversion to malonyl-CoA. Plastidic ACCase is thought to be a key regulatory enzyme in plant lipid metabolism (Somerville et al., 2000). Within the plastid, acetyl-CoA and malonyl-CoA are used in a series of repetitive reactions catalyzed by seven enzymes of the fatty acid synthase complex (Somerville et al., 2000; Stumpf, 1987). As shown in Figure 15.1, the fatty acid synthase complex catalyzes the building or elongation of fatty acids (in the form of acyl-ACPs) two carbon units at a time. The first cycle involves a condensation reaction where ketoacyl-ACP synthase III catalyzes the attachment of the acyl carboxyl group of the elongating acyl-ACP to the methylene carbon of an incoming malonyl-CoA with the resulting loss of the malonyl C_3 carboxyl as CO_2 (reaction 3). The resulting C_3 keto group of the extended acyl-ACP chain is then removed in three steps (reactions 4, 5, and 6). The process is repeated six times using a different isozyme of ketoacyl-ACP synthase (synthase I, enzyme 7) to produce palmitoyl-ACP (16:0-ACP). Palmitoyl-ACP is elongated to 18:0-ACP by a similar mechanism involving the catalytic action of a third isozyme of ketoacyl-ACP synthase (synthase II, enzyme 8). Finally, 18:0-ACP is desaturated to 18:1-ACP by the catalytic action of stearoyl-ACP Δ^9-desaturase.

Newly synthesized acyl-ACPs are the initial sources of the acyl groups required for glycerolipid assembly in both the plastid and extraplasitidic compartments (the endoplasmic reticulum and mitochondria). Export of fatty acids from the plastid requires that the acyl group of acyl-ACPs first be released from ACP as the free or unesterified fatty acid. This is accomplished by one or more chain length specific thioesterases within the plastid. Subsequently, free fatty acids (FFAs) are then converted to the acyl-CoA derivative and released to the cytosol via acyl-CoA synthase situated in the plastid envelope (Koo et al., 2004). These acyl-CoAs can be used for glycerolipid biosynthesis by either the mitochondria or the endoplasmic reticulum, and in the latter case is referred to as the "eukaryotic pathway" for lipid biosynthesis (Somerville et al., 2000). In contrast, within the plastid, acyl-ACPs are used directly for glycerolipid biosynthesis by what is commonly referred to as the "prokaryotic pathway." Except for the first acyl transferase reaction, which occurs in the stroma phase, glycerolipid assembly in the plastid occurs almost entirely in the inner membrane of the plastid envelope, the name given to the double membrane that characteristically delimits plastids (Somerville et al., 2000; Figure 15.2).

Glycerolipid assembly commences with the sequential acylation of glycerol-3-phosphate to form lyso-phosphatidic acid (LPA) and then phosphatidic acid (PA). These reactions are catalyzed by glycerol-3-phosphate and LPA acyltransferases, respectively. Each of these acyltranferases is somewhat specific for the fatty acids that they insert into positions 1 and 2 of glycerol-3-phosphate. The first acyltransferase prefers 18:1-ACP while the second prefers 16:0-ACP. (The specificities of the corresponding enzymes of the eukaryotic pathway are both for 18:1-CoA.) Phosphatidic acid is a key glycerolipid intermediate in the synthesis of glycerolipids. Depending on the level or activity of plastidic PA phosphohydrolase, the diacylglycerol (DAG) moiety can be directed towards the plastidic glycolipids monogalactosyl-, digalactosyl- and sulfoquinovosyldiacylglycerol (MGDG, DGDG, SQDG) or phosphatidylglycerol (PG) (Somerville et al., 2000). It should be noted, however, that plastids contain both "prokaryotic" and "eukaryotic" molecular species of glycerolipids (especially their glycolipids). In this regard, plastids must variously rely on the cytosolic compartment for a supply of eukaryotic diacylglycerols having C_{18} fatty acids in both the *sn*-1 and 2 positions.

The relative contributions of the prokaryotic vs. eukaryotic pathways for glycerolipid assembly in plastids is somewhat species specific and is primarily manifested in the relative 16:3 vs. 18:3 fatty acid compositions of their leaf galactolipids (Jamieson and Reid, 1971; Mongrand et al., 1998). Plants such as *Arabidopsis,* spinach and tobacco contain high amounts of 16:3 and 18:3, and are thought to utilize both the prokaryotic and eukaryotic pathways for galactolipid assembly. They are loosely referred to as "16:3 plants." Similarly, plants such as corn, wheat and pea contain only 18:3 in their galactolipids and are thought to exclusively use the eukaryotic pathway for galactolipid assembly. These plants are thus known as "18:3 plants" (Mongrand et al., 1998). The mechanism whereby these eukaryotic DAGs gain entry into the plastid and enter plastidic lipid metabolism remains poorly understood and is thought to somehow involve phosphatidylcholine (PC) of the endoplasmic reticulum. Early work suggested that the entire DAG moiety of PC was incorporated into plastidic glycolipids (Roughan and Slack, 1982). However, more recent studies indicate that extraplastidic lyso-PC is acylated to form plastidic PC and the latter then serves as the source of eukaryotic diacylglycerol (Mongrand et al., 2000). Whatever the case, the precise mechanism whereby plastidic or extraplastidic PC gives rise to the diacylglycerol moiety required for eukaryotic glycerolipid assembly in plastids remains unknown.

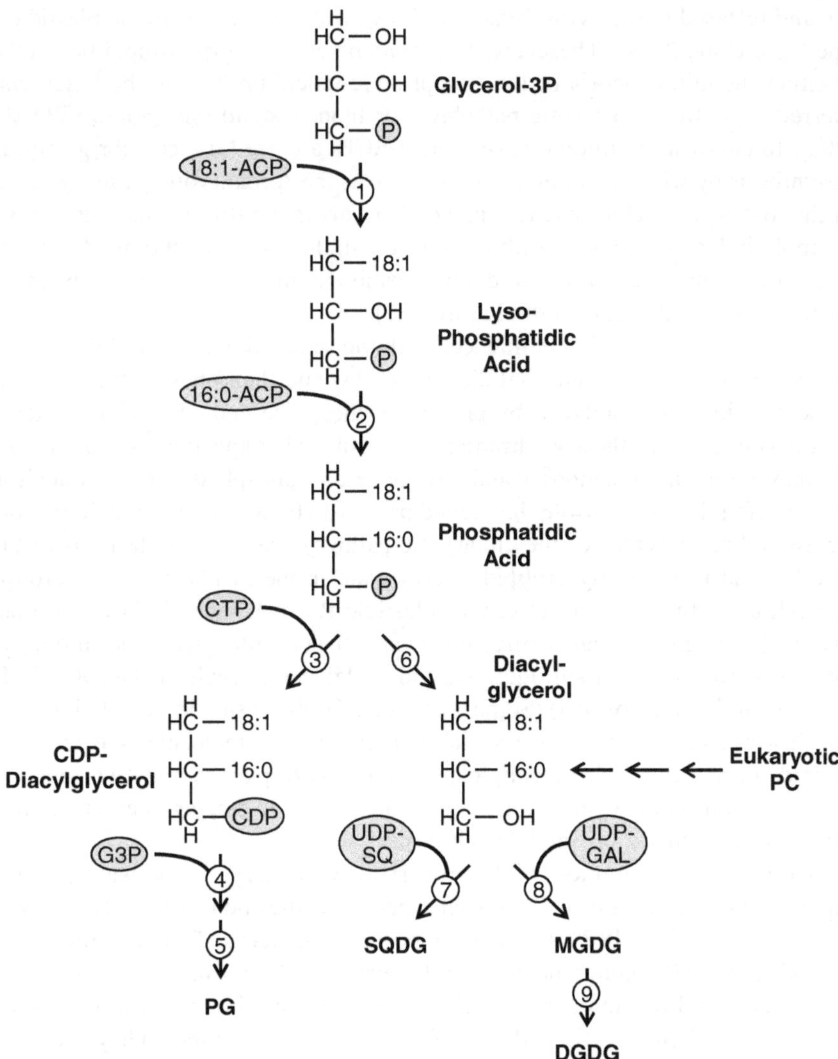

FIGURE 15.2 Enzymes and reactions of glycerolipid assembly in plastids. 1 = glycerol-3-phosphate acyltransferase; 2 = lyso-phosphatidic acid acyltransferase; 3 = CDP-diacylglycerol cytidyltransferase; 4 = phosphatidylglycerol-phosphate synthase; 5 = phosphatidylglycerol-phosphate phosphatase; 6 = phosphatidic acid phosphatase; 7 = UDP-sulfoquinovose:DAG quinovosyltransferase; 8 = UDP-galactose:DAG galactosyltransferase; 9 = galactolipid galactosyltransferase.

15.3 PLASTIDS AS A MODEL FOR TEACHING ABOUT PLANT LIPID METABOLISM

With over 50 years of relatively intensive research, a solid foundation of scientific information about plant lipid metabolism has accumulated. The metabolic pathways and the organelles involved, especially the central role that plastids play, have essentially all been defined. The cofactor requirements for individual reactions and entire metabolic processes have been fully characterized, and the elucidation of the biochemical and genetic modes of regulation is well under way (Beisson et al., 2003; Browse and Somerville, 1991; Hobbs et al., 2004; Ohlrogge and Browse, 1995). The isolation of plastids from a variety of sources is now relatively routine, and they can be assayed under a variety of *in vitro* conditions that emulate a number of *in vivo* situations of plant lipid metabolism. Most importantly, they can be manipulated in a manner that will have a predictable impact on lipid metabolism. As a result, we are now in an excellent position to use plastids as a model system in a student-based experimental approach to illustrate many of the principles and techniques of plant lipid metabolism suitable for college-level classroom or laboratory instruction. We have used many of these in our own teaching activities. Several are described in greater detail in the pages that follow.

15.4 ISOLATION OF PLASTIDS AND PLASTID ENVELOPES

Plastids have been isolated from a variety of plant tissues including leaves, roots, flowers, fruits, seeds and tubers (Price et al., 1987). Choice of plastid/tissue type and isolation protocol both largely depend on the objectives or requirements for each line of investigation. Because plastids (especially chloroplasts) are relatively large and fragile organelles, they are very sensitive to the shear forces as well as abrupt changes in tonicity and hydrostatic pressure that are potentially generated during tissue disruption. Thus, the main isolation criteria, besides yield and purity, are intactness and ease of isolation. Two favorite plastids for studies of lipid metabolism are chloroplasts from spinach leaves and leucoplasts from pea roots. This is primarily because these plants are relatively easily grown, and a relatively good yield of high quality plastids is easily obtained from these tissues. More importantly, these two plastid types represent two extremes in specialized or functional forms, each with a markedly different basis for elucidating the complexity of metabolic interactions that necessarily must occur. In general, the best results are still obtained using a combination of differential centrifugation techniques combined with density gradient purification with Percoll silica sol (Price et al., 1987).

Once purified plastids have been isolated, they can become the starting material for the isolation of plastid envelopes when more detailed localization studies are required. The isolation of chloroplasts, leucoplasts and envelopes typically requires markedly different protocols. Two separate protocols that we have used successfully for intact chloroplasts and leucoplasts are summarized below. These are followed by a protocol for isolation of chloroplast envelope membranes.

15.4.1 Spinach Leaf Chloroplasts

The method for isolation and purification of chloroplasts described here is based on the combined protocols of Nakatani and Barber (1977) and Joy and Mills (1987), which entail first the isolation of chloroplasts by differential centrifugation in low ionic strength media followed by purification of intact chloroplasts on a step gradient of Percoll. Briefly, 20–25 g of fresh, rapidly expanding spinach leaves are homogenized in 100 mL of ice-cold buffer containing 0.33 M sorbitol, 0.2 mM $MgCl_2$ and 20 mM MES (pH 6.5). Homogenization is achieved with two to three 5-second bursts at maximum speed in a chilled Waring blender. Although this does not completely homogenize all of the leaf tissue, it minimizes shear force damage to the chloroplasts.

The homogenate is then filtered through six to eight layers of cheesecloth by gravity only (i.e., without wringing the cheesecloth) and the filtrate is centrifuged at 2200 × *g* for 30 sec (allowing a maximum of 60 sec for rotor acceleration and deceleration). The 2200 × *g* supernatant (still somewhat green) is then decanted and discarded. Some of the soft surface of the 2200 × *g* pellet is easily lost at this step, which is permissible because this contains mostly ruptured or damaged chloroplasts. The remaining pellet is gently resuspended in a volume of cold resuspension buffer equivalent to the 2200 × *g* supernatant and containing 0.33 M sorbitol and 0.5 mM tris-base (pH 7.5). The resuspended chloroplasts are then centrifuged a second time at 2200 × *g*, this time for 20 sec, and allowing only 40 sec for acceleration and deceleration. As before, the 2200 × *g* supernatant is carefully discarded, this time being careful to not lose any of the pellet, and the resulting pellet is gently resuspended in 4.0 mL of resuspension buffer. Two milliliters of the chloroplast suspension are then layered onto each of two Percoll step gradients contained in a 15 mL corex centrifuge tube and consisting of 2 mL 90%, 4 mL 40% and 4 mL 20% Percoll, with each layer containing 0.3 M sorbitol. The step gradient is finally centrifuged at 3500 × *g* for 20 min. A Sorvall SS34 rotor or equivalent works well for all of these centrifugations.

Using this procedure, intact chloroplasts sediment at the 40/90% interface while damaged chloroplasts and chloroplast membranes sediment at the 20/40% and 0/20% interfaces, respectively. After centrifugation, intact chloroplasts can then be collected by gentle aspiration with a Pasteur pipette. This procedure typically yields about 2–3 mL of chloroplasts equivalent to about 1 mg chlorophyll, which can then be used in lipid synthesis assays. It should be noted, however, that chloroplasts are extremely fragile and should be used immediately with a minimum of handling or mixing (1 h storage on ice results in loss of 25% of lipid biosynthetic capacity). Similarly, starch content in leaves immediately prior to chloroplast isolation can have a serious impact on the yield and quality of chloroplasts. Starch grains act like tiny pebbles that sediment through chloroplast membranes during centrifugation, disrupting the chloroplast structure and reducing activity. A white pellet of starch beneath the first chloroplast pellet is a telltale sign that yield and quality of chloroplasts in a given preparation will be low. Leaf starch content can be minimized by placing plants in darkness for 12 to 24 h prior to chloroplast isolation, or simply timing the harvest of leaves to occur at the very end of the growth chamber dark cycle.

15.4.2 Pea Root Leucoplasts

Pea root plastids are much smaller and seem to be less fragile than chloroplasts, but still require a fair bit of care during isolation and purification. The procedure described here is based on a protocol that we have routinely used in our laboratory

(Kleppinger-Sparace et al., 1992; Xue et al., 1997) with slight modifications from other workers (Emes and England, 1986).

Leucoplasts are isolated from aseptically grown germinating pea seeds as follows. Seeds are imbibed overnight in running tap water, and then surface sterilized for 5 min. in 5% (v/v) commercial hypochlorite solution (Chlorox®). After rinsing with sterile distilled water, pea seeds are germinated at 25°C for 5–7 days in complete darkness under sterile conditions on filter paper moistened with 5–7 mL water in Petri dishes, with 20–25 seeds per dish. Ten to 20 g of fresh root tissue (2–5 cm in length from 80 g dry seed) are thoroughly homogenized in a chilled mortar and pestle in a homogenization buffer (0.5 g tissue per mL) consisting of 50 mM Tricine (pH 7.5), 0.33 M sorbitol, 1 mM EDTA, 1 mM $MgCl_2$ and 0.1% bovine serum albumin (BSA). The homogenate is then successively filtered through two layers each of 250- and 20-μm nylon mesh to remove intact cells, nuclei and cell debris. This filtrate is centrifuged at 500 × *g* for 8 minutes to yield the plastid-enriched fraction. The crude plastids are then resuspended in 0.5 mL of homogenization buffer and purified by centrifugation through Percoll as follows. The crude plastid suspension is layered onto 5 mL of 10% (v/v) Percoll containing 50 mM Tricine buffer (pH 7.5), 0.33 M sorbitol and 0.1% (w/v) BSA in a 15 mL corex centrifuge tube, which is then centrifuged at 4000 × *g* for 5 min. Plastids pelleted through Percoll are washed to remove residual Percoll and BSA by resuspending them in 5.0 mL of 1.0 mM bis-tris-propane buffer (pH 7.5) containing 0.33 M sorbitol and centrifuging the suspension at 1000 × *g* for 5 min. Washed plastids are finally resuspended in 1 mL the same buffer to yield a plastid protein concentration of approximately 1 mg/mL, which can then be used in plastid lipid metabolism assays.

15.4.3 Isolation of Chloroplast Envelopes

For those students of plant lipid metabolism who require more precise localization or enzymatic studies, purified plastids can be further manipulated to isolate the envelope membranes. Historically, envelope membranes have been prepared mainly from spinach chloroplasts (Douce et al., 1973) and pea chloroplasts (Cline et al., 1981). However, with considerable interest and effort shifting to the model plant *Arabidopsis thaliana*, it is appropriate that we include a protocol for the isolation of chloroplasts and chloroplast envelopes from this plant.

The method we describe here is based on the combined protocols of Awai et al. (2001) and Ferro et al. (2003), with some important differences from the plastid isolation procedures described earlier. *Arabidopsis* plants are grown at 23°C for 3–4 weeks with an 8–12 h light cycle (light intensity approximately 150 $\mu Mol \cdot m^{-2} \cdot s^{-1}$), and then stored in the dark for one light cycle prior to chloroplast isolation to minimize the starch content, as previously noted. Again, it is important to keep all solutions and to perform all manipulations at 0–4°C. Four hundred to 500 g of *A. thaliana* leaves are homogenized in 2–3 volumes of homogenization buffer (0.33 M sucrose, 30 mM NaPPi, 10 mM MOPS-KOH, pH 7.8, 0.1% defatted BSA) for two 3-sec bursts, then filtered through a single layer of Miracloth.

For chloroplast enrichment the filtrate is centrifuged for 8 min at 1200 × *g* using slow braking for large volume (100–250 mL) centrifuge containers. The initial pellets are then resuspended in (2–5 mL each) resuspension media (0.33 M sucrose, 10 mM MOPS-KOH, pH 7.8) then combined and brought up to 200–250 mL volume and centrifuged in smaller centrifuge tubes for 2 min at 1200 × *g*, with slow braking. These pellets are gently resuspended in minimal (1–2 mL added/pellet) resuspension media prior to layering approximately 5 mL over each of 6–12 premade Percoll step gradients (preferably 6 × 30 mL volume) kept at 0°C. The gradients are each composed of 3 mL 55% Percoll, 10 mL 35% Percoll and 10 mL 20% Percoll, with each layer containing 0.33 M sucrose, 10 mM MOPS-KOH, pH 7.8. After centrifuging the gradients at 3000 × *g* for 20 min at 2°C followed by slow braking, the intact chloroplasts are removed from the 55/35% interface and diluted sevenfold with resuspension buffer prior to centrifuging for 5 min at 3000 × *g* and slow braking, to obtain a pellet of highly purified intact chloroplasts.

Once obtained, purified intact chloroplasts are first lysed by gently resuspending each pellet in minimal (2–4 mL per original Percoll gradient) hypotonic medium (10 mM MOPS-KOH, pH 7.8 containing, as needed, 1 mM benzamidine, 1 mM phenylmethylsulfonyl fluoride and 0.5 mM ε-amino caproic acid). Note that the hypotonic media is essentially the same as the resuspension buffer described above but lacks the osmoticum sucrose. It also contains protease inhibitors for labile enzyme assays or for samples intended for protein gels that are to be run later. Envelope membranes are next purified by layering the lysate over 2–4 discontinuous sucrose step gradients (prepared ahead of time but kept refrigerated). Each sucrose gradient is composed of 2 mL 0.93 M, 3mL 0.6 M, 4 mL 0.3 M sucrose, each layer buffered with 10 mM MOPS-NaOH, pH 7.8 and containing protease inhibitors as necessary. Envelope membranes are purified from this lysate by centrifuging the sucrose step gradients at 70,000 × *g* for 1 h using an appropriate rotor (i.e., a swinging bucket rotor like the Beckman SW41-Ti rotor, Beckman, Urbana, IL, or vertical rotor). Envelope membranes are collected at the 0.93/0.6 M sucrose interface, then concentrated by diluting three- to fourfold with the hypotonic buffer containing protease inhibitors, then pelleting at 110,000 × *g* for 1 h (Beckman SW41-Ti rotor) (Awai et al., 2001; Ferro et al., 2003).

15.5 ASSAY CONDITIONS FOR PLASTID LIPID BIOSYNTHESIS

Given the fact that plastids are the site of *de novo* fatty acid biosynthesis, and that they can use newly synthesized fatty acids for their own glycerolipid assembly, reaction mixtures designed to promote all combined aspects of lipid biosynthesis as illustrated in Figures 15.1 and 15.2 can be relatively complex. Table 15.1 summarizes the optimum concentrations of the various cofactors known to be required for fatty acid and glycerolipid biosynthesis in spinach chloroplasts and pea root leucoplasts. Depending on the experimental objectives, specific aspects of lipid metabolism in these plastids can be monitored by the addition of an assortment of radiolabeled precursors. These commonly include 0.5 to 2.0 μCi of either ^{14}C-acetate or ^{14}C-glycerol-3-phosphate, which will label the acyl or glycerol backbone moieties, respectively, of many key intermediates and products of plastidic lipid metabolism. Alternatively, more specific radiolabeled precursors such as $Na_2{}^{35}SO_4$ or UDP-^{14}C-galactose can

TABLE 15.1
Concentrations of Various Cofactors Required for Lipid Biosynthesis from Acetate and Glycerol-3-Phosphate by Spinach Chloroplasts and Pea Root Leucoplasts

Reagent	Acetate Incorporation into Chloroplast Lipids[1]	Acetate Incorporation into Pea Root Leucoplast Lipids[2]	G3P Incorporation into Pea Root Leucoplast Lipids[3]
		Concentration (mM)	
Osmoticants			
Sorbitol	0.3	—	0.31
Sucrose or Sorbitol	—	0.31	—
Buffers			
Tricine (pH 7.9)	33.0	—	—
Bis-Tris-Propane (pH 8.0)	—	100.0	—
Bis-Tris-Propane (pH 7.5)	—	—	100.0
Lipid Substrates			
Na-acetate	0.15	0.20	0.20
DL-glycerol-3-phosphate	0.40	1.00	0.16
$KHCO_3$	10.0	15.0	15.0
Lipid Biosynthesis Cofactors			
ATP	1.50	6.00	2.00
Coenzyme A	0.20	0.20	0.05
NADH	—	1.00	0.50
NADPH	—	1.00	0.50
$MgCl_2$	2.00	6.00	2.00
$MnCl_2$	—	1.00	—
KH_2PO_4	0.20	—	—
Plastids, Reaction Volume, Incubation Time and Temperature			
Chlorophyll (μg)	50–200		
Protein (μg)		40–60	40–60
Volume (mL)	1.0	0.5	0.5
Time/Temperature (h/°C)	1.0/25	1.0/25	1.0/25

[1] Mudd et al. (1987); [2]Stahl and Sparace (1991); [3]Xue et al. (1997)

be used for more precise studies of sulfolipid (Kleppinger-Sparace et al., 1985) or galactolipid biosynthesis (Douce, 1974; Heinz and Roughan, 1983; McCune, 1995).

Reactions are typically started by the addition of an aliquot of the appropriate plastid preparation, and then incubated at 25°C for various time intervals as required. A major difference between the incubation conditions required by chloroplasts vs. leucoplasts is light. As mentioned earlier, lipid biosynthesis in chloroplasts is coupled to photosynthesis with little or no biosynthesis occurring in darkness (Table 15.2). The best results are obtained when high-intensity light (and incubation) is provided with a photosynthetic Warburg apparatus. Other illumination systems with appropriate shielding from heat from incandescent lights can also be used. Reactions are finally terminated by the addition of a monophasic organic chloroform/methanol mixture designed to extract lipids and then partition them into chloroform (Bligh and Dyer, 1959; Mudd and DeZacks, 1981). Aliquots of the chloroform phase are finally analyzed by a variety of standard techniques of lipidology, including scintillation counting, thin layer chromatography and gas-liquid chromatography. A thin-layer chromatography system that we have found particularly useful for resolving radiolabeled lipids synthesized by plastids is illustrated in Figure 15.3 (Sparace and Mudd, 1982). Under the standard reaction conditions

TABLE 15.2

The Effects of Cofactor Deletion or Addition on the Rates of Lipid Biosynthesis from Acetate and Glycerol-3-Phosphate by Spinach Chloroplasts and Pea Root Leucoplasts

	Rate (% of Control)		
Treatment	**Acetate Incorporation into Chloroplast Lipids[1]**	**Acetate Incorporation into Pea Root Leucoplast Lipids[2]**	**G3P Incorporation into Pea Root Leucoplast Lipids[3]**
Control	100	100	100
Dark	11	—	—
Boiled	—[a]	< 1	—
– Sorbitol	32	40[b]	—
– $KHCO_3$	25	18	55
– CoA	104	3	5
– G3P	57	82	—
– Acetate	—	—	77
– ATP	58	1	5
– ATP + CTP (3 mM)	72	—	—
– NADH	—	42	74
– NADPH	—	107	92
– NADH, – NADPH	—	19	67
– $MgCl_2$	23	43	25
+ $MnCl_2$ (2 mM)	69	—	—
+ Lyso-PC (10 μM)	147	—	48
+ UDP-Galactose (0.2 mM)	125	—	120
+ Triton X-100 (0.001, 0.01%)	—	115, 6	121[c]
+ β-Mercaptoethanol (1 mM)	—	—	4
+ Dithiothreitol (1 mM)	—	53	2

[1] Sparace and Mudd, 1982; [2]Stahl, 1990; [3]Xue, 1993; [a] "–" = not determined; [b]sucrose was the osmoticant omitted; [c]Triton X-100 was 0.008% for glycerol-3-phosphate incorporation.

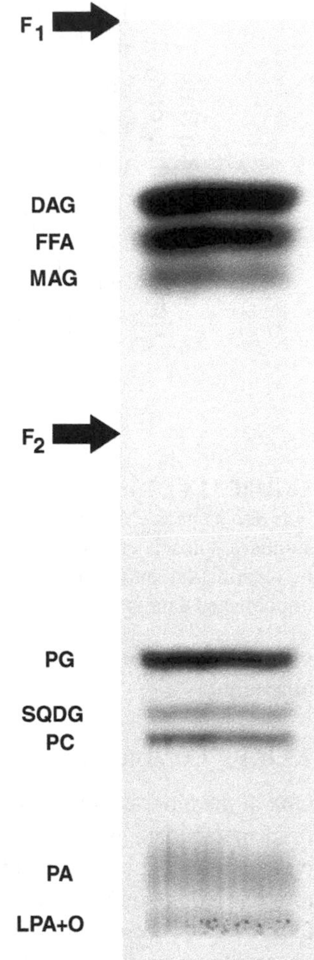

FIGURE 15.3 Autoradiogram of thin layer chromatographic separation of radiolabeled spinach chloroplast lipids synthesized from ^{14}C-acetate. Lipid extracts are applied to pre-run Brinkman Polygram Sil G TLC plates (0.25 mm × 20 cm × 20 cm) that are first developed with acetone/acetic acid/water (100:2:1, v/v) until the solvent front (F_1) reaches the top of the plate. After thorough drying under flowing N_2, TLC plates are then developed with chloroform/methanol/NH_4OH/H_2O (65/35/2/2, v/v) until the solvent front (F_2) moves approximately 12 cm up the plate. DAG = diacylglycerol; FFA = free fatty acids; MAG = monoacylglycerol; PG = phosphatidylglycerol; SQDG = sulfoquinovosyldiacylglycerol; PC = phosphatidylcholine; PA = phosphatidic acid; LPA+O = lyso-phosphatidic acid plus origin. (Reproduced from Sparace and Mudd, 1982, with permission.)

given above, lipid biosynthesis from these precursors is linear for about 30 min with spinach chloroplasts (Figure 15.4) and for up to 6 h with pea root leucoplasts (Figure 15.5) (S.A. Sparace, unpublished observations; Stahl, 1990; Xue, 1993). The shorter period of linearity for lipid biosynthesis in chloroplasts is very likely due to the great fragility of chloroplasts, as noted earlier. However, depending on the reaction vessel used and whether gentle agitation is provided, we have also observed that chloroplasts will eventually sediment out during prolonged incubation periods. This can cause localized depletion of externally supplied cofactors.

15.6 USING PLASTIDS FOR EDUCATIONAL PURPOSES IN PLANT LIPID METABOLISM

Once high-quality (primarily intact) plastids (or envelope membranes) have been obtained, they can be used in a variety of applications or activities suitable for college-level instruction emphasizing plant lipid metabolism. We have used many

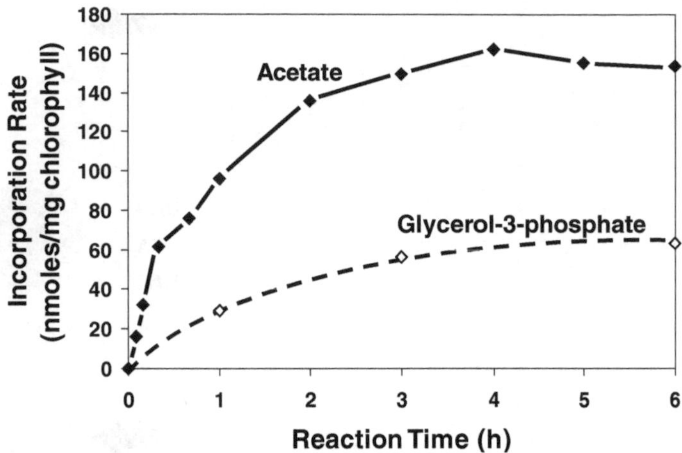

FIGURE 15.4 Timecourse study for ^{14}C-acetate vs. ^{14}C-glycerol-3-phosphate incorporation into spinach chloroplast lipids. Different chloroplast preparations were incubated under standard spinach chloroplast reaction conditions for increasing lengths of time followed by termination, extraction and quantification of total radioactivity in lipids. (S.A. Sparace, unpublished data.)

of these in our own teaching activities. Here we describe a few that are most amenable.

15.6.1 Cofactor Requirements for Plastid Lipid Biosynsynthesis

One of the most straightforward applications of plastids as a tool for teaching about lipid metabolism is to determine (or confirm) the cofactor requirements for lipid biosynthesis by plastids. Students can observe the effects of omitting various cofactors one by one on the rates of lipid biosynthesis from radioactive precursors under the standard optimized conditions described earlier. They can then relate their observations to the well-characterized pathways of lipid metabolism illustrated in Figures 15.1 and 15.2. Typical data sets for such experiments with spinach chloroplasts and pea root plastids are shown in Table 15.2. Students can then speculate on whether plastids can provide (or synthesize) any of their own cofactors, based on their metabolic capabilities for photosynthesis glycolysis and/or the pentose phosphate pathway.

Similarly, the effects of adding extra cofactors or regulators such as unlabeled UDP-galactose and dithiothreitol to the standard reaction mixtures can also be assessed, and then students can suggest possible interpretations or mechanisms for their observations. For example, lipid biosynthesis in pea root leucoplasts is completely dependent on exogenously supplied ATP, whereas chloroplasts are only partly dependent on externally supplied ATP. The difference is due to the generation of ATP during photosynthesis in chloroplasts. This simple fact underscores the experimental requirement for light as described earlier when working with chloroplasts (as opposed to leucoplasts).

Another notable observation in Table 15.3 is that lipid biosynthesis from acetate is much more dependent on exogenously supplied reduced nucleotides (NADH

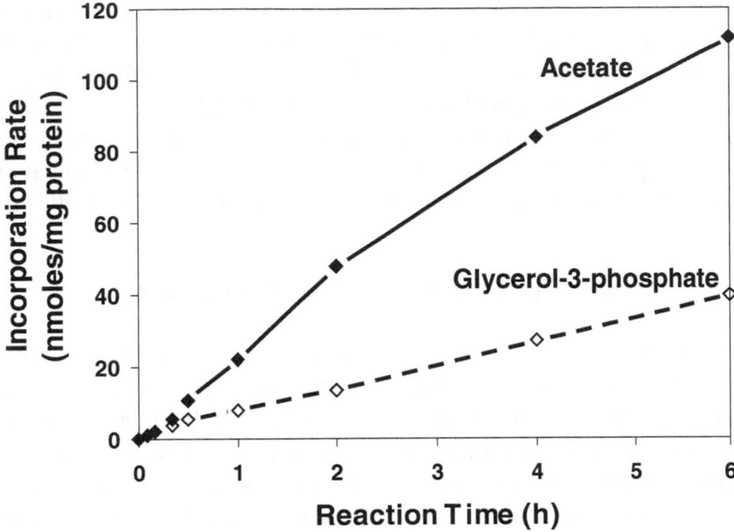

FIGURE 15.5 Timecourse study for ^{14}C-acetate vs. ^{14}C-glycerol-3-phosphate incorporation into pea root leucoplast lipids. Different plastid preparations were incubated under standard pea root plastid reaction conditions for increasing lengths of time followed by termination, extraction and quantification of total radioactivity in lipids. (Modified from Stahl, 1990 and Xue, 1993.)

and NADPH) than lipid biosynthesis from glycerol-3-phosphate in pea root leucoplasts. This difference echoes the reduced nucleotide requirements for *de novo* fatty acid biosynthesis, but not for glycerolipid biosynthesis. It also implies that isolated pea root leucoplasts have a small endogenous supply of acyl moieties for some glycerolipid assembly, but that some *de novo* fatty acid synthesis is also required. Similar studies can also be done with several other precursors of lipid metabolism that are readily utilized by plastids. These include glucose-6-phosphate, phosphoenolpyruvate, pyruvate and malate (Tetlow et al., 2005). Further, the use of these precursors will invariably change the cofactor requirements for lipid biosynthesis, especially for ATP and reduced nucleotides (Qi et al., 1995; Smith et al., 1992)

An even greater level of understanding can be probed if crude lipid extracts are resolved into their individual lipid components by thin layer chromatography (Figure 15.3) or any other separation technique. The profile of lipids synthesized by any plastid preparation is greatly dependent on the origin of the plastids, the composition of the reaction mixture and the length of time plastids are allowed to metabolize lipid precursors. For example, under standard reaction conditions, both spinach chloroplasts (Figure 15.3) and pea root leucoplasts will synthesize primarily PA, DAG, PG, PC (and triacylglycerol [TAG] in pea root leucoplasts) from radiolabeled acetate or glycerol-3-phosphate (Sparace and Mudd, 1982; Xue et al., 1997). The addition of UDP-galactose is seen in Table 15.2 to stimulate lipid biosynthesis in both chloroplasts and leucoplasts by 20–25%, suggesting a possible regulatory role for UDP-galactose. Further examination and quantification of the proportions of radioactive lipids synthesized reveals that the addition of UDP-galactose creates a "metabolic sink" where as much as 50% of the radioactivity in newly synthesized lipid is accumulated in monogalactosyldiacylglycerol with large corresponding decreases in the amounts of phosphatidic acid and diacylglycerols synthesized (Sparace and Mudd, 1982; McCune, 1995). Such simple data sets measuring the cofactor requirements for the total amounts and proportions of various lipids synthesized are replete with a variety of ways that students of lipid metabolism can exercise or demonstrate their understanding of lipid biosynthesis in plastids.

15.6.2 Timecourse Studies of Plastid Lipid Biosynthesis

Another very useful application in using plastids as a tool for learning about plant lipid metabolism is simple timecourse studies, where the effects of reaction incubation times on the proportions of various synthesized lipids are examined. With spinach chloroplasts, the proportion of radioactivity is always greatest and increasing in DAG (up to 60%), with a transient increase and then decrease in the amount of PA, and a slow steady increase in the amounts of the more polar phospholipids (Figure 15.6). In contrast, pea root leucoplasts initially accumulate most radioactivity (almost 60%) in PA within 30 min of reaction time (Figure 15.7). This is followed by a steady decline in the amounts of radioactive PA, which is accompanied by a corresponding increase in the amounts of polar phospholipids and surprisingly

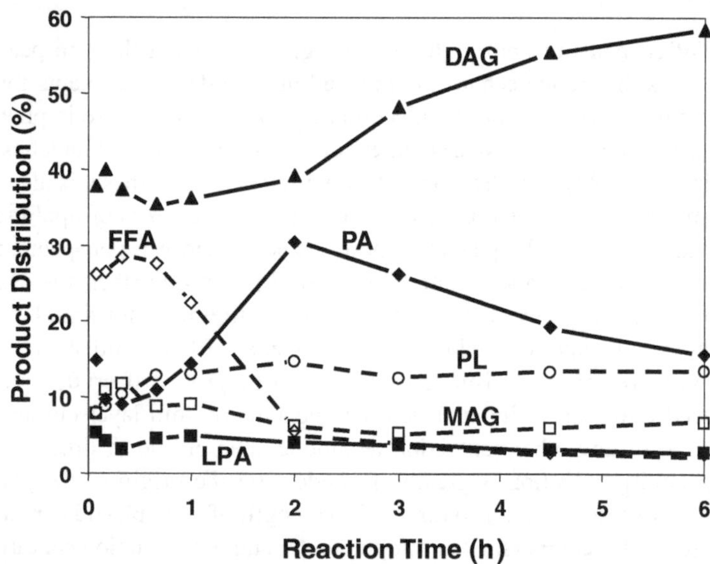

FIGURE 15.6 The effects of increasing reaction times on the distribution of radioactivity from ^{14}C-acetate among the various lipids synthesized by spinach chloroplasts. Total lipid extracts were resolved by thin layer chromatography and the amount of radioactivity in each lipid calculated as a percent of the total. DAG = diacylglycerol; MAG = monoacylglycerol; LPA = lyso-phosphatidic acid; PL = the sum of remaining radioactivity in phosphatidylglycerol, sulfoquinovosyldiacylglycerol, phosphatidylcholine and monogalactosyldiacylglycerol. (S.A. Sparace, unpublished data.)

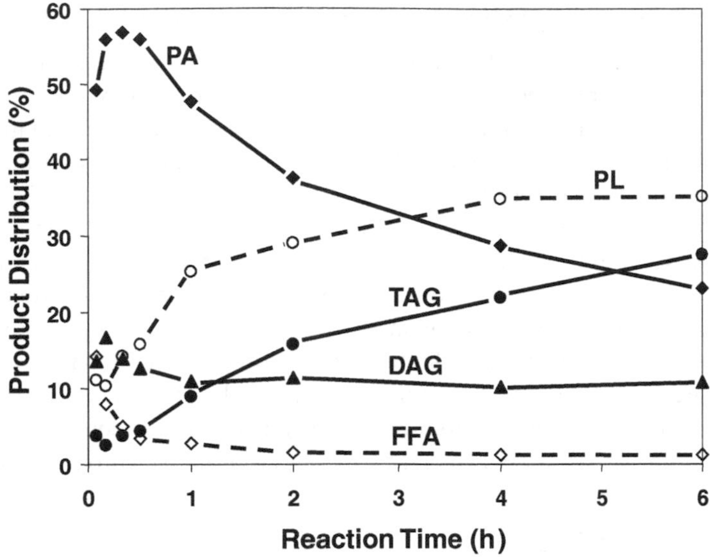

FIGURE 15.7 The effects of increasing reaction times on the distribution of radioactivity from ^{14}C-acetate among the various lipids synthesized by pea root leucoplasts. Total lipid extracts were resolved by thin layer chromatography and the amount of radioactivity in each lipid calculated as a percent of the total. TAG = triacylglycerol. (Modified from Stahl, 1990.)

substantial amounts (up to 27%) of radioactivity accumulated in TAG. These differences, especially between the amounts of PA vs. DAG accumulated by spinach chloroplasts vs. pea root leucoplasts, is clearly related to the physiological nature of these two plant species. Spinach and pea are classic examples of "16:3" and "18:3" plants, respectively. As mentioned earlier, an important difference between these two groups of plants is the relative activity of plastidic PA phosphatase in each species. This enzyme has a higher activity in 16:3 plants (like spinach), resulting in larger amounts of DAG accumulated *in vitro* by plastids actively synthesizing lipid, whereas in 18:3 plants (like pea) the activity of the enzyme is greatly suppressed, thus accounting for the accumulation of high amounts of PA. These patterns of lipid metabolism have essentially become dogma of plant lipid biochemistry.

Finally, as mentioned earlier, the profiles of lipid accumulation can be modified by manipulating the components of the reaction mixture. Besides the addition of UDP-galactose as mentioned earlier, some interesting manipulations are the addition of detergents which can disrupt membrane structure and inhibit plastidic lipid biosynthesis (Mudd and DeZacks, 1981; Stahl, 1990), the addition of lyso-phosphatidylcholine, which promotes phosphatidylcholine synthesis in both spinach chloroplasts (Sparace and Mudd, 1982) and leek chloroplasts (Bessoule et al., 1995), the addition CTP and Mn^{2+} which promote phosphatidylglycerol synthesis in spinach chloroplasts (Mudd and DeZacks, 1981; Mudd et al., 1987; Sparace and Mudd, 1982) and the adjustment of incubation pH which seems to primarily affect the activities of stearate desaturase (Stahl and Sparace; 1991) and phosphatidic acid phosphatase (Xue et al., 1997) in pea root plastids.

15.6.3 Cytosolic Interactions and Metabolite Uptake Studies

As described earlier, plastids interact greatly with the cytosolic compartment via a number of different membrane transporters in order to carry out their cellular responsibilities. Some of these can be coupled directly to plastidic lipid metabolism in a meaningful way to probe their role in plastidic lipid biosynthesis. In this regard, we have used what has become known as the triose-phosphate shuttle (Figure 15.8) to promote the generation of an internal supply of plastidic ATP. This shuttle mechanism thus bypasses the light requirement for sulfolipid biosynthesis in spinach chloroplasts (Kleppinger-Sparace and Mudd, 1987) or the exogenously supplied ATP for *de novo* fatty acid biosynthesis in pea root leucoplasts (Kleppinger-Sparace et al., 1992). The triose-phosphate shuttle mechanism, which relies on both the triose-phosphate and dicarboxylate transporters, promotes the uptake of dihydroxyacetone phosphate and its metabolism to 3-phosphoglycerate via the coupled activities of triose-phosphate isomerase, glyceraldehyde-3-phosphate dehydrogenase and phos-

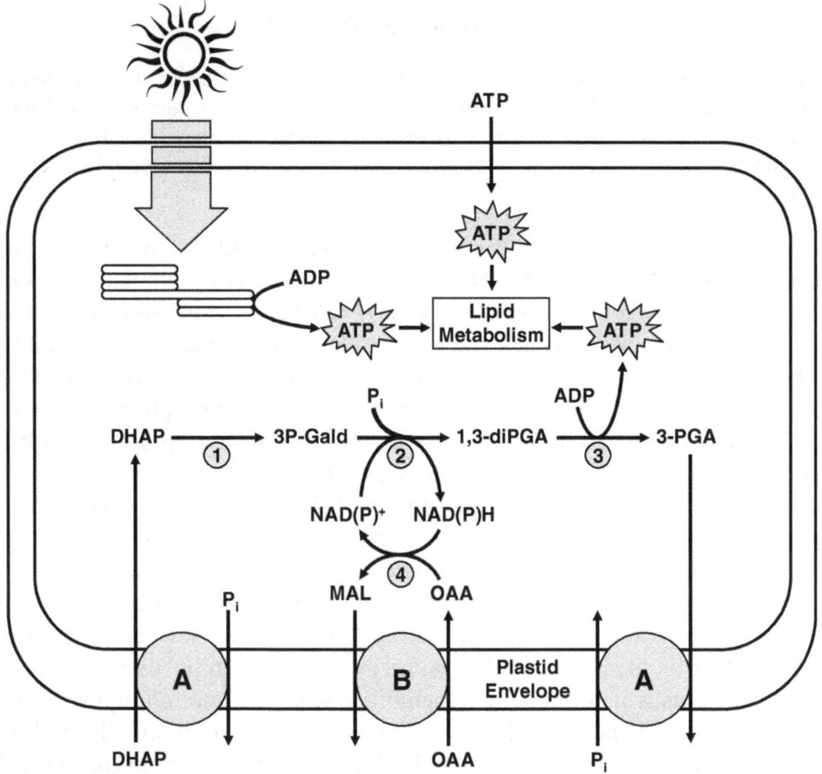

FIGURE 15.8 The triose-phosphate shuttle for the generation of intraplastidic ATP. A = triose-phosphate translocator; B = dicarboxylate translocator; 1 = triose-phosphate isomerase; 2 = glyceraldehyde-3-phosphate dehydrogenase; 3 = phosphoglycerate kinase; 4 = malate dehydrogenase; DHAP = dihydroxyacetone phosphate; 3P-Gald = 3-phosphoglyceraldehyde; 1,3-diPGA = 1,3-diphosphoglyceric acid; 3-PGA = 3-phosphoglyceric acid; MAL = malic acid; OAA = oxaloacetic acid. (Modified from Kleppinger-Sparace et al., 1992.)

phoglycerate kinase. The latter enzyme becomes the physiological source of the ATP required for lipid biosynthesis. With a little imagination, similar studies for either teaching or research purposes can be designed to examine the effects of the uptake and metabolism of other key intermediates such as glucose-6-phosphate and phosphoenolpyruvate on plastidic lipid metabolism (Tetlow et al., 2005).

15.6.4 Envelope-Based Studies of Plastidic Lipid Metabolism

Although the preparation of chloroplast envelope membranes requires considerably more time and effort than the isolation of chloroplasts, a higher degree of experimental precision or focus can be had by using envelopes instead of whole chloroplasts. As mentioned earlier, the chloroplast envelope is engaged in glycerolipid assembly, which can proceed in the absence of *de novo* fatty acid biosynthesis. Depending on the source of the chloroplasts, envelope membranes are capable of carrying out essentially all enzymic reactions shown in Figure 15.2 except the very first.

The enzyme responsible for the first reaction (glycerol-3-phosphate acyltransferase) is a soluble plastid enzyme that is lost during the isolation of envelope membranes (Frentzen et al., 1983; Joyard and Douce, 1977). Thus, to accurately monitor glycerolipid assembly in envelope membranes, the appropriate molecular species of lyso-PA and suitable acyl donors must be used. More specifically, oleoyl-lyso-PA and palmitoyl-ACP are physiologically the most relevant and preferred precursors for glycerolipid assembly. In the latter case, palmitoyl-CoA can also be used in place of the acyl-ACP, although the latter is preferred by the enzyme (Frentzen et al., 1983).

Unfortunately, these precursors tend to be either very costly or not commercially available (especially when radiolabeled). Thus, they are often synthesized by those investigators who require their use. This adds a significant element of complexity or difficulty in performing such experiments. Nevertheless, when the appropriate precursors are available, a number of interesting experiments can be performed.

In this regard, in a series of elegant experiments, Andrews and Mudd (1985; Mudd et al., 1987) were able to carefully monitor the flow of radioactivity from ^{14}C-palmitoyl-ACP into PA, CDP-diacylglycerol, PG-phosphate and ultimately the final product, PG, in the envelope membranes from pea chloroplasts. Similarly, using spinach chloroplast envelopes in combination with an extract of chloroplast proteins, Joyard and Douce (1977) were able to demonstrate the incorporation of ^{14}C-glycerol-3-phosphate into lyso-PA, PA, DAG and then into MGDG upon the addition of UDP-galactose. The reader is referred to the original works of these researchers for additional details on the compositions of reaction mixtures and assay conditions for this line of experiments.

15.7 SUMMARY AND CONCLUSIONS

Plastids occupy a central role in the overall physiology of plants, especially here in regard to plant lipid metabolism. Over the past 50 years or so, lipid scientists have learned a great deal about the metabolic processes of plastids that either support or are directly involved in *de novo* lipid biosynthesis. Many of the approaches for the isolation of plastids and their use in studies of plant lipid metabolism have become fairly routine and almost standardized with many predictable patterns of lipid metabolism, and we acknowledge the many contributions of all plant lipid scientists in this area. The accumulation of all of this information makes plastids an excellent model to use in teaching about plant lipid metabolism. We expect that plastids will continue to be an important tool for both teaching and research in plant lipid biochemistry, and we wish students and researchers all the best in their studies of plant lipid metabolism.

REFERENCES

Andrews, J., Mudd, J.B. Phosphatidylglycerol synthesis in pea chloroplasts: Pathway and localization, *Plant Physiol.* 79, 259–265, 1985.

Awai, K., Maréchal, E., Block, M.A., Brun, D., Masuda, T., Shimada, H., Takamiya, K.-I., Ohta, H., Joyard, J., Two types of MGDG synthase genes, found widely in both 16:3 and 18:3 plants, differentially mediate galactolipid syntheses in photosynthetic and nonphotosynthetic tissues in *Arabidopsis thaliana*, *PNAS* 98, 10960–10965, 2001.

Bao, X., Focke, M., Pollard, M., Ohlrogge, J., Understanding *in vivo* carbon precursor supply for fatty acid synthesis in leaf tissue, *Plant J.*, 22, 39–50, 2000.

Beisson, F., Koo, A., Ruuska, S., Schwender, J., Pollard, M., Thelen, J., Paddock, T., Salas, J., Savage, L., Milcamps, A., Mahske, V.B., Cho, Y., Ohlrogge, J.B., Arabidopsis genes involved in acyl lipid metabolism. A 2003 census of the candidates, a study of the distribution of expressed sequence tags in organs, and a Web-based database, *Plant Physiol.* 132, 681–697, 2003.

Bessoule, J.J., Testet, E., Cassagne, C., Synthesis of phosphatidylcholine in the chloroplast envelope after import of lysophosphatidylcholine from endoplasmic reticulum membranes, *Eur. J. Biochem.* 228, 490–497, 1995.

Bligh, E.G., Dyer, W.J., A rapid method of total lipid extraction and purification, *Can. J. Biochem. Physiol.* 37, 911–917, 1959.

Browse, J., Somerville, C., Glycerolipid metabolism, biochemistry and regulation, *Ann. Rev. Plant Physiol. Plant Mol. Biol.* 42, 467–506, 1991.

Cline, K., Andrews, J., Mersey, B., Newcomb, E.H., Keegstra, K., Separation and characterization of inner and outer envelope membranes of pea chloroplasts, *PNAS* 78, 3595–3599, 1981.

Crawford, N.M., Kahn, M.L., Leustek, T., Long, S.R., Nitrogen and sulfur, in *Biochemistry and Molecular Biology of Plants,* Buchanan, B.B., Gruissem, W., Jones, R.L., eds., American Society of Plant Physiologists, Rockville, MD, chap. 16, 2000.

Douce, R., Site of biosynthesis of galactolipids in spinach chloroplasts, *Science,* 183, 852–853, 1974.

Douce, R., Holtz, B., Benson, A.A., Isolation and properties of the envelope of spinach chloroplasts, *J. Biol.Chem.* 248, 7215–7222, 1973.

Emes, M.J., England, S., Purification of plastids from higher-plant roots, *Planta* 168, 161–166, 1986.

Ferro, M., Salvi, D., Brugiere, S., Miras, S., Kowalski, S., Louwagie, M., Garin, J., Joyard, J., Rolland, N., Proteomics of the chloroplast envelope membranes from *Arabidopsis thaliana, Mol. Cell Proteomics* 2, 325–345, 2003.

Frentzen, M., Heinz, E., McKeon, T.A., Stumpf, P.K., Specificities and selectivities of glycerol-3-phosphate acyltransferase and monoacylglycerol-3-phosphate acyltransferase from pea and spinach chloroplasts, *Eur. J. Biochem.* 129, 629–636, 1983.

Halliwell, B., *Chloroplast Metabolism: The Structure and Function of Chloroplasts in Green Leaf Cells.* Oxford University Press, New York, 1981.

Heinz, E., Roughan, P.G., Similarities and differences in lipid metabolism of chloroplasts isolated from 16:3 and 18:3 plants, *Plant Physiol.* 72, 273–279, 1983.

Hobbs, D.H., Flintham, J.E., Hills, M.J., Genetic control of storage oil synthesis in seeds of Arabidopsis, *Plant Physiol.* 136, 3341–3349, 2004.

Jamieson, G.R., Reid, E.H., The occurrence of hexadeca-7,10,13-trienoic acid in the leaf lipids of angiosperms, *Phytochem.* 10, 1837–1843, 1971.

Joy, K.W., Mills, W.R., Purification of chloroplasts using silica sols, in *Methods in Enzymology* 148 (*Plant Cell Membranes)*, Packer, L., Douce, R., eds., Academic Press, San Diego, 179–188, 1987.

Joyard, J., Douce, R., Site of synthesis of phosphatidic acid and diacylglycerol in spinach chloroplasts, *Biochim. Biophys. Acta* 486, 273–285, 1977.

Joyard, J., Teyssier, E., Miège, Berny-Seigneurin, D., Maréchal, E., Block, M.A., Dorne, A.-J., Rolland, N., Ajlani, G., Douce, R., The biochemical machinery of plastid envelope membranes, *Plant Physiol.* 118, 715–723, 1998.

Kleppinger-Sparace, K.F., Mudd, J.B., Biosynthesis of sulfoquinovosyldiacylglycerol in higher plants: The incorporation of $^{35}SO_4$ by intact chloroplasts in darkness, *Plant Physiol.* 84, 682–287, 1987.

Kleppinger-Sparace, K.F., Mudd, J.B., Bishop, D.J., Biosynthesis of sulfoquinovosyldiacylglycerol in higher plants: The incorporation of $^{35}SO_4$ by intact chloroplasts, *Arch. Biochem. Biophys.*, 240, 859–865, 1985.

Kleppinger-Sparace, K.F., Stahl, R.J., Sparace, S.A., Energy requirements for fatty acid and glycerolipid biosynthesis from acetate by isolated pea root plastids, *Plant Physiol.* 98, 723–727, 1992.

Koo, A.J.K., Ohlrogge, J.B., Pollard, M., On the export of fatty acids from the chloroplast, *J. Biol. Chem.* 279, 16101–16110, 2004.

Kuhn, D.N., Knauf, K., Stumpf, P.K., Subcellular localization of acetyl-CoA synthetase in leaf protoplasts of *Spinacia oleracea, Arch. Biochem. Biophysics* 209, 441–450, 1981.

Malkin, R., Niyogi, K., Photosynthesis, in *Biochemistry and Molecular Biology of Plants,* Buchanan, B.B., Gruissem, W., Jones, R.L., eds., American Society of Plant Physiologists, Rockville, MD, chap. 12, 2000.

McCune, L.M., Characterization of galactolipid synthesis in pea root plastids, M.S. thesis, McGill University, Montreal, 1995.

Mongrand, S., Bessoule, J.-J., Cabantous, F., Cassagne, C., The C16:3/C18:3 fatty acid balance in photosynthetic tissues from 468 plant species, *Phytochem.* 49, 1049–1064, 1998.

Mongrand, S., Cassagne, C., Bessoule, J.-J., Import of Lyso-Phosphatidylcholine into chloroplasts likely at the origin of eukaryotic plastidial lipids, *Plant Physiol.* 122, 845–852, 2000.

Mudd, J.B., Andrews, J.E., Sparace, S.A., Phosphatidylglycerol synthesis in chloroplast membranes, in *Methods in Enzymology* 148 (*Plant Cell Membranes)*, Packer, L., Douce, R., eds., Academic Press, San Diego, 338–345, 1987.

Mudd, J.B., DeZacks, R., Synthesis of phosphatidylglycerol by chloroplasts from leaves of *Spinacia oleracea* L. (spinach), *Arch. Biochem. Biophys.* 209, 584–591.

Nakatani, H.Y., Barber, J., An improved method for isolating chloroplasts retaining their outer membranes, *Biochim. Biophys. Acta* 461, 510–512, 1977.

Neuhaus, H.E., Emes, M.J., Nonphotosynthetic metabolism in plastids, *Ann. Rev. Plant Physiol. Plant Mol. Biol.* 51: 111–140, 2000.

Ohlrogge, J., Browse, J., Lipid biosynthesis, *The Plant Cell,* 7, 957–970, 1995.

Price, C.A., Cushman, J.C., Mendiola-Morgenthaler, L.R., Reardon, E.M., Isolation of plastids in density gradients of percoll and other silica sols, in *Methods in Enzymology* 148 *(Plant Cell Membranes)*, Packer, L., Douce, R., eds., Academic Press, San Diego, 157–179, 1987.

Qi, Q., Kleppinger-Sparace, K.F., Sparace, S.A., The utilization of glycolytic intermediates as precursors for fatty acid and glycerolipid biosynthesis in pea root plastids, *Plant Physiol.* 107, 413–419, 1995.

Roughan, P.G., Slack, C.R., Cellular organization of glycerolipid metabolism, *Ann. Rev. Plant Physiol.* 33, 97–132, 1982.

Roughan, P.G., Slack, C.R., Holland, R., High rates of [1-^{14}C]acetate incorporation into the lipid of isolated chloroplasts, *Biochem. J.* 158, 593–601, 1976.

Slack, C.R., Roughan, P.G., The kinetics of incorporation *in vivo* of [1-^{14}C]acetate and [1-^{14}C] carbon dioxide into the fatty acids of glycerolipids in developing leaves, *Biochem. J.* 152, 217–228, 1975.

Smith, R.G., Gauthier, D.A., Dennis, D.T., Turpin, D.H., Malate- and pyruvate-dependent fatty acid synthesis in leucoplasts from developing castor endosperm, *Plant Physiol.* 98, 1233–1238, 1992.

Somerville, C., Browse, J., Jaworski, J.G., Ohlrogge, J.B., Lipids, in *Biochemistry and Molecular Biology of Plants,* Buchanan, B.B., Gruissem, W., Jones, R.L., eds., American Society of Plant Physiologists, Rockville, MD, chap. 10, 2000.

Sparace, S.A., Mudd, J.B., Studies on chloroplast lipid metabolism: Stimulation of phosphatidylglycerol biosynthesis and analysis of the radioactive lipid, in *Biochemistry and Metabolism of Plant Lipids,* Wintermans, J.F.G.M., Kuiper, P.J.C., eds., Elsevier Biomedical Press, Amsterdam, 111–119, 1982.

Staehelin, L.A., Newcomb, E.H., Membrane structure and membranous organelles, in *Biochemistry and Molecular Biology of Plants,* Buchanan, B.B., Gruissem, W., Jones, R.L., Eds., American Society of Plant Physiologists, Rockville, MD, chap. 1, 2000.

Stahl, R.J., Fatty acid and glycerolipid biosynthesis in pea root plastids, M.S. thesis, McGill University, Montreal, 1990.

Stahl, R.J., Sparace, S.A., Characterization of fatty acid synthesis in isolated pea root plastids, *Plant Physiol.* 96, 602–608, 1991.

Stumpf, P.K., The biosynthesis of saturated fatty acids, in *The Biochemistry of Plants,* Vol. 9, Stumpf, P.K., Conn, E.E., eds., Academic Press, New York, 121–136, 1987.

Stumpf, P.K., Barber, G.A., Fat metabolism in plants IX—Enzymic synthesis of long chain fatty acids by avocado particles, *J. Biol. Chem.* 227, 407–417, 1957.

Tetlow, I.J., Rawsthorne, S., Raines, C., Emes, M.J., Plastid metabolic pathways, in *Plastids,* Møller, S.G., Ed., CRC Press, Boca Raton FL, 2005, chap. 3.

Waters, M., Pyke, K., Plastid development and differentiation, in *Plastids,* Møller, S.G., Ed., CRC Press, Boca Raton, FL, chap. 2, 2005.

Weber, A.P.M., Schwacke, R., Flügge, U.-I., Solute transporters of the plastid envelope membrane, *Ann. Rev. Plant Biol.* 56, 133–164, 2005.

Xue, L., Glycerolipid biosynthesis in pea root plastids, M.S. thesis, McGill University, Montreal, 1993.

Xue, L., McCune, L.M., Kleppinger-Sparace, K.F., Brown, M.J., Pomeroy, M.K., Sparace, S.A., Characterization of the glycerolipid composition and biosynthetic capacity of pea root plastids, *Plant Physiol.* 113, 549–557, 1997.

Index

A

B

C

G

H

I

J

K

L

M

N

O

P

Q

R

S

T

U

V

W

Y

Z

Milton Keynes UK
Ingram Content Group UK Ltd.
UKHW040110071024
449327UK00019B/946

9 780367 388225